CARCERAL GEOG

T0231582

The 'punitive turn' has brought about new ways of thinking about geography and the state, and has highlighted spaces of incarceration as a new terrain for exploration by geographers. Carceral geography offers a geographical perspective on incarceration, and this volume accordingly tracks the ideas, practices and engagements that have shaped the development of this new and vibrant subdiscipline, and scopes out future research directions. By conveying a sense of the debates, directions, and threads within the field of carceral geography, it traces the inner workings of this dynamic field, its synergies with criminology and prison sociology, and its likely future trajectories. Synthesizing existing work in carceral geography, and exploring the future directions it might take, the book develops a notion of the 'carceral' as spatial, emplaced, mobile, embodied and affective.

This book is dedicated to Maggie, and to Margaret.

Carceral Geography
Spaces and Practices of Incarceration

DOMINIQUE MORAN
University of Birmingham, UK

Routledge
Taylor & Francis Group

LONDON AND NEW YORK

First published 2015 by Ashgate Publishing

Published 2016 by Routledge
2 Park Square, Milton Park, Abingdon, Oxon OX14 4RN
711 Third Avenue, New York, NY 10017, USA

First issued in paperback 2017

Routledge is an imprint of the Taylor & Francis Group, an informa business

British Library Cataloguing in Publication Data
A catalogue record for this book is available from the British Library

The Library of Congress has cataloged the printed edition as follows:
Moran, Dominique.
  Carceral geography : spaces and practices of incarceration / by Dominique Moran.
    page cm
  Includes bibliographical references and index.
  ISBN 978-1-4094-5234-8 (hardback)
  1. Imprisonment--Social aspects. 2. Prisons. I. Title.
  HV8705.M668 2015
  365--dc23

                                                          2014031731

ISBN 13: 978-1-138-30846-6 (pbk)
ISBN 13: 978-1-4094-5234-8 (hbk)

Contents

List of Figures   *vii*
About the Author   *ix*
Acknowledgements   *xi*

1   Introduction   1

2   Origins and Dialogues   7

PART I CARCERAL SPACE

3   Carceral Space   17

4   The Emotional and Embodied Geographies of Prison Life   29

5   Carceral TimeSpace   43

PART II GEOGRAPHIES OF CARCERAL SYSTEMS

6   Geographies of Carceral Systems   59

7   Prison Transport and Disciplined Mobility   71

8   Inside/Outside and the Contested Prison Boundary   87

PART III THE CARCERAL AND A PUNITIVE STATE

9   The Carceral and a Punitive State   105

10   Prison Buildings and the Design of Carceral Space   113

11   Carceral Cultural Landscapes, Post-Prisons and
     the Spectacle of Punishment   129

12   Afterword   149

Bibliography   *153*
Index   *183*

List of Figures

7.1   Prisoner transport rail coach preserved at the 'Stasi Prison',
      Gedenkstätte Berlin-Hohenschönhausen, former East Berlin   82

11.1 Merchandizing penal history at Katajanokka Hotel, Helsinki   133
11.2 Merchandizing penal history at Langholmen Hotel, Stockholm   133
11.3 Utilising carceral symbolism at Langholmen Hotel, Stockholm   134
11.4 Patarei prison, Tallinn, Estonia   136
11.5 Art installation at Patarei prison, Tallinn, Estonia   136
11.6 The 'Stasi Prison', at Hohenschönhausen, former East Berlin   140
11.7 Preserved 'interview room' at the 'Stasi Prison',
      Hohenschönhausen, former East Berlin   140

About the Author

Dominique Moran is Reader in Carceral Geography at the School of Geography, Earth and Environmental Sciences, University of Birmingham, UK. Her research in carceral geography explores geographical perspectives on incarceration. She currently holds over £1m of funding from the UK Economic and Social Research Council, for research into prison visitation and recidivism, and prison design, and she is an editor of the recent collection *Carceral Spaces: Mobility and Agency in Imprisonment and Migrant Detention* (Ashgate 2013). Her work is transdisciplinary, informed by and extending theoretical developments in geography, criminology and prison sociology, and also interfacing with contemporary debates over hyperincarceration, recidivism and the advance of the punitive state. She publishes in leading journals including *Transactions of the Institute of British Geographers* and *Environment and Planning D: Society and Space*.

Acknowledgements

This book is the result of a variety of events, processes and discussions, in a variety of contexts and with a multitude of individuals, for all of which, and to all of whom, I am grateful.

Key to the development of carceral geography within my own work has been the support of the UK Economic and Social Research Council (ESRC) to three separate research projects and a seminar series. All of this work is collaborative, much of it across disciplinary boundaries, and the resulting dialogues have been productive, challenging, and insightful. First, *Women in the Russian Penal System: The Role of Distance in the Theory and Practice of Imprisonment in Late Soviet and Post-Soviet Russia* (RES-062-23-0026, 2006–2010), undertaken with Judith Pallot and Laura Piacentini, and more recently *Breaking the Cycle? Prison Visitation and Recidivism in the UK* (ES/K002023/1, 2012–2016), with Louise Dixon and Marie Hutton; *'Fear-suffused Environments' or Potential to Rehabilitate? Prison Architecture, Design and Technology and the Lived Experience of Carceral Spaces* (ES/K011081/1, 2014–2016), with Yvonne Jewkes and Jennifer Turner, and *Exploring Everyday Practice and Resistance in Immigration Detention* (ESRC Seminar Series 2012–2014), led by Nick Gill, with Alex Hall, Imogen Tyler and Mary Bosworth. Although the empirical results of these projects are published elsewhere, it is the transdisciplinary dialogues they stimulated which led to the ideas behind this book.

Underpinned by this support, dialogue around carceral geography has developed through numerous themed conference sessions, where the process of co-organising, chairing and discussing opened new areas for debate. These included: at the Annual Meetings of the Association of American Geographers in Washington, DC in 2010; Seattle, WA in 2011 (nine sessions co-organised with Matt Mitchelson, Lauren Martin, Andrew Burridge, Jenna Loyd, Nancy Hiemstra, Geoffrey Boyce, Jennifer Ridgley and Hunter Jackson); Los Angeles, CA in 2013 (with Shaul Cohen); and Tampa, FL in 2014 (with Karen Morin); and at the Annual Conferences of the Royal Geographical Society with the Institute of British Geographers in London in 2010; Edinburgh in 2012 (with Jon Coaffee); and London in 2014 (with Anna Schliehe). Conference sessions also led to the publication of the edited volume *Carceral Spaces: Mobility and Agency in Imprisonment and Migrant Detention* (2013) co-edited with Nick Gill and Deirdre Conlon.

A Visiting Fellowship at the Aleksanteri Institute, University of Helsinki, in summer of 2011, provided invaluable support for the development of this volume. The space to think and to write, fuelled by *pulla* and strong Finnish coffee, and the relaxed, scholarly environment of the Institute, enabled the ideas behind the book to take shape. Conversations with Hanna Ruutu, Anna-Liisa Heusala, Aino Saarinen, Anna Korhonen and Markku Kivinen were thought-provoking, and the opportunity to work with Anssi Keinänen and Zuzanna Bogumił opened new avenues for inquiry. Invitations, including from Lauren Martin to address colleagues at the University of Oulu, Department of Geography; from Ben Crewe to speak at the Prisons Research Centre, Institute of Criminology, University of Cambridge; and from Bénédicte Michalon to address the *TerrFerme* Colloquium at the *Centre National de la Recherche Scientifique* at the University of Bordeaux, France also provided welcome opportunities to discuss emergent themes. Discussions with Megan Comfort, Dan Mears, Alison Mountz, Shaul Cohen, Olivier Milhaud, Thomas Ugelvik and many, many others, have also provided considerable food for thought.

My thanks also go to Katy Crossan at Ashgate, who showed considerable forbearance when the delivery of the manuscript was repeatedly delayed by the delivery of a baby, and particularly to Shaul Cohen, who stepped in to chair AAG sessions in Los Angeles when it became clear just how ambitious the plan to travel with a newborn really was …

Dominique Moran
Birmingham

Chapter 1
Introduction

The 'punitive turn' has brought about new ways of thinking about geography and the state, and has highlighted spaces of incarceration as a new terrain for exploration by geographers. Carceral geography offers a geographical perspective on incarceration, and this volume accordingly tracks the ideas, practices and engagements that have shaped the development of this new and vibrant subdiscipline, and scopes out future research directions. By conveying a sense of the debates, directions, and threads within the field of carceral geography, it traces the inner workings of this dynamic field, its synergies with criminology and prison sociology, and its likely future trajectories. Synthesising existing work in carceral geography, and exploring the future directions it might take, the book develops a notion of the 'carceral' as spatial, emplaced, mobile, embodied and affective.

This introduction defines carceral geography as an emergent subdiscipline of human geography, provides a brief overview of the three emerging themes within geographical scholarship of incarceration around which the book is structured, and sets the scene for the following chapter which traces the origins of work in this field and its dialogues with research in other disciplinary areas.

Carceral Geography

Carceral geography is a new, fast-moving and fast-developing sub-discipline. Although the first paper by a geographer published squarely in this field was probably the work of Teresa Dirsuweit (1999) on women's imprisonment in South Africa, the enormous potential of spaces of incarceration for geographical enquiry was highlighted by Chris Philo (2001), who turned a review of Kantrowitz' (1996) *Close Control: Managing a Maximum Security Prison; The Story of Ragen's Stateville Penitentiary* into an agenda-setting article germinating the ideas which have informed the early development of this area of research, in terms of a critical engagement with spaces of confinement and a dialogue with the work of Foucault. The early work of Ruth Wilson Gilmore (1999, 2002) also ignited geographers' interest in prisons as a 'project of state-building' (ibid. 2002: 16). More than a decade after Philo's 2001 paper was published, and with the sub-discipline proving an increasingly vibrant field of scholarship, this book now provides an overview and synthesis, and suggests some future directions for research.

The term 'carceral geography' (Moran et al. 2011) was coined to describe this vibrant field of geographical research into practices of incarceration, viewing such carceral spaces broadly as a type of institution whose distributional geographies, and geographies of internal and external social and spatial relations, should be explored. Although the first geographical studies of spaces of confinement were individual, disparate works in different contexts and with different aims and theoretical framings, the gradual coalescence of a corpus of work in this field has come to reflect a growing dialogue with the work of Goffman (1961) on the 'total institution', of Foucault (1979) on the development of the prison, surveillance, and the regulation of space and docility of bodies, and of Agamben (1998, 2005) on the notion of a space of exception, where sovereign power suspends the law, producing a zone of abandonment. Although the ideas advanced by Goffman, Foucault and Agamben often underpin this scholarship, other theoretical frames are also increasingly deployed, for example de Certeau's concept of tactics (Baer 2005), along with theories of liminality and mobility (Moran et al. 2011).

Reflecting upon the emergence of carceral geography, Philo (2012: 4) described it as a sub-strand of 'geographical security studies', drawing attention to consideration of 'the spaces set aside for 'securing' – detaining, locking up/away – problematic populations of one kind or another', but there is an argument for a slightly more nuanced interpretation of the work in this emerging field.

There are three related and interconnected emergent themes within carceral geography, which may be broadly conceived of as the nature of carceral spaces and experiences within them, the spatial geographies of carceral systems, and the relationship between the carceral and an increasingly punitive state. This book takes as its structuring logic these three thematic areas, surveying existing scholarship, both within human geography and within allied disciplines which have taken an overtly spatial focus, and suggesting directions in which future scholarship could move.

Structure

Following this introduction, the book unfolds first through a chapter which outlines the origins of carceral geography and its on-going dialogues with disciplines with a long-standing engagement with spaces and practices of incarceration. It then proceeds through three sections, first on carceral space, next the spatial geographies of carceral systems, and finally the relationship between the 'carceral' and a punitive state. These three sections are by no means discrete, and there is significant overlap between the issues addressed in each of them; structuring the book in this way suggests neither that existing work is restricted to these themes, nor that they should necessarily shape the ways in which future scholarship develops. Their utility here is in structuring discussion of the state of the field at present, and in enabling connections to be made between the themes emergent within carceral geography, and contemporary theory building within human geography more broadly.

Each of the three sections comprises three chapters, the first of which provides an overview of existing work, followed by two which develop and take forward specific areas of work. In the first section, on carceral space, the first chapter surveys many of the initial geographical studies of incarceration which can be characterised as developing an interest in the nature and experience of carceral spaces, and in which theorisations of imprisonment informed by Foucault have been debated and contested. Dirsuweit (1999) examined a prison for women in South Africa, showing that rather than being rendering 'docile', prisoner resistance to omni-disciplinary control was expressed through the reclaiming of culturally-defined prison space. In New Mexico, Sibley and van Hoven also contested this Foucauldian regulation of prison space and the docility of bodies, describing 'spaces ... produced and reproduced on a daily basis' (van Hoven and Sibley 2008: 1016), and the agency of inmates making 'their own spaces, material and imagined' (Sibley and van Hoven 2008: 205). In the UK, Baer (2005) identified the personalisation of prison space, suggesting that this spatial modification reflected the construction of the meaning of prison spaces. Informed by understandings of carceral space as lived and experienced, and of the integral relationship between space and time (Massey 2005), the two chapters which follow suggest that carceral geographers could usefully consider the embodied experience of imprisonment, and could highlight the temporal aspect of imprisonment. These two chapters accordingly take up these two themes, exploring first the corporeal nature of incarceration, and then the experience of what might be termed carceral 'TimeSpace'.

The second section focuses on spatial geographies of incarceration, one step removed from the spaces of individual institutions. Research into the geographical distribution of sites of incarceration across space has often been inspired by concern for the impact of the siting of places of incarceration on the communities which host or surround them, and has frequently considered critiques and reinterpretations of the 'total institution' Goffman (1961). Examples have included Che (2005) on the location of a prison in Appalachian Pennsylvania, US; Glasmeier and Farrigan (2007) on impacts of prison development in persistently poor rural places in the US; Engel's (2007) research on prison location in the American Midwest; studies of the effects of 'geographies of punishment' on experiences of incarceration (e.g. Moran et al. 2011, Pallot 2007); Bonds' (2009) questioning of prison siting as a means of encouraging economic development, and Mitchelson (2012) on spatial interdependencies between prisons and cities in Georgia, US. Much of this work extends critiques of the 'total institution' (Goffman 1961), and suggests that the 'carceral' is something much more than merely the spaces in which individuals are confined. In considering how carceral geographers could further advance this field of inquiry, the following two chapters identify and develop themes emerging from this work, which are implicit in the operation of carceral systems for both the communities which host them, the prisoners who inhabit them, and those who visit the incarcerated. These two chapters accordingly focus first on theorisations of disciplined, coerced, or governmental mobility, as an inherent part of the functionality of carceral systems, and next on the nature of the boundary between what is considered 'inside' and 'outside' of the prison, drawing attention to these two aspects of carceral systems in ways which resonate with contemporary themes within critical human geography such as mobility, affect and embodiment.

Finally, the third section explores the notion of the 'carceral' as a social construction relevant both within and outside physical spaces of incarceration, which has informed some of the most recent research into the relationship between the 'carceral' and a punitive state. The first chapter provides context, detailing the 'new punitiveness' (Pratt et al. 2011), and its relationship to hyperincarceration and the carceral 'churn' (Peck and Theodore 2009: 251). The work of social theorists and geographers such as Wacquant (2010a, b and c), Gilmore (2007) and Peck and Theodore (2009), has called for greater attention to the causes of and solutions to hyperincarceration (Wacquant 2010b: 74) 'prisonfare' (Wacquant 2010c: 197), and the carceral churn. At higher level of abstraction again from lived spaces of incarceration, Peck (2003) and Peck and Theodore (2009) have discussed the relationship between prisons and the metropolis in the context of hyperincarceration, in the aftermath of what Wacquant (2011a: 3) described as 'a brutal swing from the social to the penal management of poverty' particularly in the United States, with a 'punitive revamping' of public policy tackling urban marginality, and establishing a 'single carceral continuum' between the ghetto and the prison (Wacquant 2000: 384). This conceptualisation of the prison as a locus on the carceral continuum resonates with the work of Baer and Ravneberg (2008) who problematised the conceptualisation of a binary distinction between 'inside' and 'outside', instead positing prisons as 'heterotopic spaces outside of and different from other spaces, but still inside the general social order' (ibid. 2008: 214).

In taking forward this scholarship, the following two chapters draw attention to the built form of prisons, as 'different spaces', occupying a variety of roles within the social order. The first discusses their design and functionality, arguing that they can be understood as the physical manifestation of penal philosophies, and arguing that carceral geographers could usefully turn their attention to prison design as a means of understanding what it is that prisons are intended to do, and the ways in which they achieve this through the deployment of space and architecture. The second chapter takes forward the idea of a carceral cultural landscape, and advances the notion of the 'post-prison', a site no longer functioning as a space of incarceration, but nevertheless still saturated with, and arguably communicative of, messages about the purpose of imprisonment both in terms of the of the system during which it was constructed, and during which it is protected, conserved, demolished, or left to decay.

Context and Comparison

Carceral geography has emerged at a specific moment both with reference to geographical theory-building and in relation to current developments in penal policies; namely the rise of the 'new punitiveness' and the advance of increasingly rapacious carceral systems. In surveying the emergence of carceral geography, this book draws on existing research both within and outside of geography, detailing the themes which characterise this work and the approaches taken to understanding the spaces and practices of imprisonment. The picture of carceral geography which results is, therefore, coloured significantly by the nature of existing scholarship, and the contexts in which research has been carried out. Much research has been undertaken in the United States; with its context of mass incarceration inevitably compelling academic attention, and the book reflects this emphasis. Although the carceral *milieu* and practices of the United States are argued to be exported to other contexts, with increased punitivity arguably becoming a characteristic of much of the Anglophone world (e.g. Pratt et al. 2011), research drawing on other, different, carceral contexts, such as the Russian Federation, and Scandinavia, is also discussed here. The intention is not to provide a comparison; rather to identify themes and issues which have commonality and purchase across contexts, albeit expressed or experienced differently according to local contingencies.

The imperative to shed light on the US carceral system has within human geography developed in tandem with an overt abolitionist perspective (e.g. Loyd et al. 2012), drawing together perspectives on incarceration and migrant detention to understand these practices as 'state violence' and to argue that 'freedom of movement and freedom to inhabit are necessarily connected' (ibid. 8–10). Whilst entirely consonant with this sentiment, this book addresses a different set of intentions, and whilst recognising that there is much to be gained from a synthesis of studies of both 'mainstream' imprisonment and migrant detention (e.g. Moran et al. 2013a), it focuses primarily, although not exclusively, on imprisonment rather than migrant detention. The intention here is to open a space for consideration of the *purposes* of imprisonment and the contexts in which these are constructed – a theme well developed within criminology and prison sociology, but as yet largely elided within carceral geography. Turner (2013a: 35) has argued that carceral geography has tended to 'position spaces of imprisonment within thematics of containment and exclusion, which removes from consideration the particular contextual issues of reform and rehabilitation'. This book seeks to question and to problematize the notions of reform and rehabilitation as they apply to contemporary practices of imprisonment, within an overarching intention to draw together discourses of incarceration, and to open a space for research dialogues which are dynamically open to transdisciplinarity, which are both informed by and extend theoretical developments in geography, but which also, and critically, interface with contemporary debates over hyperincarceration, recidivism and the advance of the punitive state.

Chapter 2
Origins and Dialogues

Carceral geography sits at a nexus of interrelated developments in geographical research: the immense influence of the engagement between Michel Foucault and questions of space, place and geography; the prominence within contemporary critical human geography of the ideas of Giorgio Agamben about 'bare life' and spaces of exception; the growing currency of the work of Loïc Wacquant on hyperincarceration and the punitive turn in the United States and Western Europe, and the integration of these perspectives into human geography by scholars such as Ruth Wilson Gilmore and Jamie Peck. In parallel, the recent spatial turn within criminology and prison sociology (Kindynis 2014) increasingly draws upon human geographical understandings of space and spatiality as multiplicitous and heterogeneous, lived and experienced (e.g. Pickering 2014). As a result, an increasingly interdisciplinary approach has emerged within the social sciences which has opened a space for the coalescence of work under the heading of carceral geography (Moran et al. 2011, Moran 2013a).

This coalescence has occurred at a critical moment in contemporary penal practice, with the expansion of 'workfare' and 'prisonfare' policies, the criminalisation of immigration and the expansion of the carceral estate. Of particular note in relation to studies within criminology and prison sociology, is that carceral geography's focus on the spaces and practices of confinement extends both to 'mainstream' incarceration, i.e. of individuals detained by the prevailing legal system and migrant detention, where irregular migrants and 'refused' asylum seekers are detained, ostensibly pending decisions on admittance or repatriation (e.g. Moran et al. 2013a).

The coalescence of carceral geography is a relatively recent development with human geography, but it both builds on longstanding human geographical concerns, and engages with scholarship in disciplines for which incarceration and confinement have long been core research themes. The emergence of carceral geography at this particular juncture, though, is indicative of its connections to the wider body of work in critical human geography, and of its synergies with the praxis of social and political change aimed at challenging and transforming prevalent relations, systems, and structures of inequality and oppression. By way of providing context for the subsequent sections of the book, this chapter therefore briefly traces the origins and emergence of carceral geography, locates it in relation to dialogues with cognate disciplines such as criminology and prison sociology, and considers the socio-political context in which geographers have turned their attention to spaces and practices of incarceration. It proceeds by first discussing carceral geography in relation to themes within contemporary human geography, and its connections to criminology and prison sociology, and finally by engaging with the contemporary discourses of punitiveness and hyperincarceration.

Human Geography and Carceral Geography

Attempting to 'fix' the relative position of one area of research within the fluid context of contemporary human geography is a futile undertaking, not only in terms of its achievability given the fast-moving nature of this scholarship, but also in terms of the risk of imposing any sense of boundedness on the scope of this work and its potential for fruitful convergences. The purpose of this section of the chapter, then, is to *describe*, rather than to *prescribe*, the nature of the subdiscipline of carceral geography in terms of its recent emergence, and some of the synergies already apparent with more 'established' themes of work in human geography.

Starting first with terminology, the notion of the 'carceral' in relation to space arguably came first into the vocabulary of human geography through Davis' (1990) *City of Quartz*, in which he examined the hardening of the cityscape of Los Angeles, calling it a 'carceral city' with a pervasive security agenda which enmeshed the city in networks of surveillance, where police lobbying for expanded law-and-order land use threatened to 'convert an entire salient of Downtown-East Los Angeles into a vast penal colony' (Davis 2006: 254), and in which he observed at the time of writing that the prison population was already the highest in the nation, looking set to double within the next decade. Davis' (1990) notion of the 'carceral city' had an immediate influence, turning geographers' attention towards gated communities, video observation cameras, and the securitisation of urban space. Whilst existing in close relationship to these securitised spaces as potential destinations for the criminalised underclasses excluded from affluent 'forbidden cities', prisons at this time themselves remained largely under-researched by human geographers, despite the astonishing acceleration of the expansion of the US', and especially California's, prison estate (Gilmore 2007).

The *metaphor* of the prison, though, via the notion of the Panopticon (after Foucault 1979), gained increased purchase in discourses relating to the changing nature of urban space. Similarities between this 'ideal prison' and urban space began to be discussed in relation to unavoidable visibility via surveillance (Cohen 1985, Soja 1989, Hannah 1997, Koskela 2000), where urban citizens' constant awareness of their own visibility was argued to recall Foucault's notion of 'conscious and permanent visibility that assures the automatic functioning of power' (Foucault 1979: 201). Similarly, the 'unverifiability' of urban surveillance (Koskela 2000: 253), never knowing whether cameras are active or whether their footage is being watched, was seen to have parallels with the inmate's lack of knowledge about when exactly (s)he is being observed; 'the inmate must never know whether he is being looked at any moment, but he must be sure that he may always be so' (Foucault 1979: 201). The anonymity of power both in the idealised Panopticon and in urban space, and the absence of force which Panoptic surveillance is argued to enable through the power of the 'gaze', were further axes of similarity.

Whilst the 'ideal prison', thus understood via the principle of the Panopticon, orbited these discourses of securitisation and surveillance of urban space, arguably the first foray by a geographer writing in the English language into a 'really existing' prison was undertaken in South Africa by Dirsuweit (1999).[1] And from the outset, as exemplified by her work, geographers' engagement with carceral spaces has sought both to question the literal interpretation of Foucault's work on the Panopticon, and to situate prisons in their spatial and cultural context, emphasising the connections between imprisonment and other forms of confinement. Although the 'central ambition' of Dirsuweit's (1999) study was to 'provide a map of a carceral institution' and 'a description of the institution as a space of omni-disciplinary control' (ibid. 82), interestingly the entry point into this study was the decades-long history of South African study of the 'compound', a tightly controlled, basic barracks, built and used by mining companies to house an all-male black migrant work force, and understood as an urban carceral environment. The functionality of surveillance within the compound as a means of imposing industrial discipline and control over the work force, meshed with racialized criminal laws and institutional controls to see it emerge as a new kind of urban spatial structure in South Africa. Underpinning Dirsuweit's study, therefore, was a concern for the ways in which the constitution of prisoner identity, and the codes of conduct which disciplined prisoner behaviour, could be understood in relation to previous scholarship of the compound, and the connections between workers' 'compound' and 'home' identities. Albeit a 'peculiarly South African institution' (Crush 1994: 305), which developed within a highly specific context of urban segregation, the compound as a carceral environment was integrally connected to wider spaces of identity formation and social control, and by using the compound as her 'entry point' into the prison, Dirsuweit (1999) made a strong statement about the complex interrelationships between prisons and their surrounding environments and networks.

In parallel with Dirsuweit, human geographers writing in French incorporated spaces of incarceration into their scholarship, for example with Fior (1993), Lamarre (2001), Milhaud (2009a and b) and Chantraine (2005) exploring the lived experience of spaces of detention in Switzerland, Canada and France respectively. Ricordeau and Milhaud later (2012) focused on how prison spaces are sexualised,

---

1   Earlier, unpublished geographical research into prisons includes Ferrant, A. 1997. *Containing the Crisis: Spatial Strategies and the Scottish Prison System*, unpublished PhD thesis, University of Edinburgh, Department of Geography; and Marshall, A. 1997. *Always Greener on the Other Side of the Fence? Examining the Relationship between the Built Environment, Regimes and Control in Medium Security Prisons in England and Wales*, unpublished PhD thesis, University of Birmingham, School of Geography, cited in Philo (2001).

and Touraut has addressed the collateral damage of incarceration in terms of mobility of prisoners' families and their negotiation of the prison boundary (2009, 2012).

From this earliest engagement with the prison, geographical research (in both French and English) in spaces of incarceration has shared some common understandings, whether these are made explicit or remain implicit in published work. Geographers have retained an explicit concern for and interest in, the relationship between the prison and the world 'outside', exploring this relationship in diverse ways, situating the prison in relation to its surrounding communities, and problematising the 'boundary' between inside and outside. They have also demonstrated a concern for the internal spatial arrangements of prisons, in terms of both the institutional layouts and the movements of prisoners' (and others') bodies around them in both space and time. As Philo (2001: 480) noted in his influential review of Kantrowitz' (1996) *Close Control*, the 'deceptively simple' geographies of prisons' internal arrangements are 'absolutely central to the overall workings of such carceral establishments', and studies of them have the potential to 'cast into clearer relief the many ways in which spatial systems, strategies and practices of all sorts enable power to be exercised over potentially unruly populations within relatively constrained bounds'.

These understandings, of prisons as connected rather than detached spaces, with permeable boundaries and highly significant internal geographies, arguably underpins subsequent geographical research into imprisonment. However, the prison as a site of research has been approached from a variety of angles and perspectives within human geography, using places of incarceration as a lens through which to interrogate a range of ideas and concepts, and these will be explored in later chapters.

Dialogues: Criminology and Prison Sociology

The emergence of carceral geography coincides with a spatial turn in post-Foucauldian critical prison studies (Smith 2013: 167). Traditionally, criminology and prison sociology have arguably tended towards conceptualisations of incarceration as a period of prison *time*, rather than as a space, for example through longitudinal studies of imprisonment rates, overcrowding and prisoner welfare (e.g. Jacobs and Helms 1996, Stucky et al. 2005), individual prisoners' experiences (e.g. Zamble 1992), and adjustment to incarceration (e.g. Warren et al. 2004, Thompson and Loper 2005). This tendency toward temporal interpretations perhaps stems from conceptualisations of imprisonment as a discrete period of time distinguished from the rest of prisoners' lifecourses (e.g. Pettit and Western 2004), and from the self-evident significance of variation in, and effect of, different lengths of prison sentences (Aebi and Kuhn 2000), in relation to the incarceration of people at different lifecourse stages (Cesaroni and Peterson-Badali 2005, Howse 2003, Crawley 2005, Rikard and Rosenberg 2007). However, in line with the embedding of a concern for space in response to the increased prominence of spatiality within social theory and the social sciences more broadly since the mid-to-late 1980s, these disciplines now increasingly view prison *space* as highly significant in understanding the experience of incarceration.

The growing influence of prison ethnography, outside of a US context in which Wacquant (2002) has lamented the scarcity of ethnographic research, also contributes to this concern for space. As Drake and Earle (2013: 12–13) noted, in Africa, South America, India and Europe, prison ethnographers have been 'getting close to the experiences, feelings and understandings of prison life' by accessing spaces of incarceration and observing 'telling details' of prison life which bring into sharp relief the 'meaning and essence of prison experiences and offer valuable means for understanding a little of what it really means to be imprisoned or to work in a prison'. In so doing, prison ethnographers are beginning to make explicit a longstanding implicit awareness of the significance of space which has remained underplayed in much scholarship to date. In an example of just this kind of development, Crewe et al. (2014: 56) drew attention to the 'emotional world' of the prison, arguing that rather than constituting 'environments that are unwaveringly sterile, unfailingly aggressive or emotionally undifferentiated', prisons instead have 'emotion zones', in which different emotional registers can be expressed. Crewe et al.'s (2014) paper is particularly significant in that it draws directly upon human and carceral geography in considering space and place as 'determinants of social practice and personal experience, rather than as empty theatres or neutral backcloths within and against which they occur' (ibid. 60).

With prison ethnography bringing to the fore a concern for the experience of prison spaces, another area of scholarship within criminology (e.g. Fiddler 2010, Hancock and Jewkes 2011, Jewkes 2013a) also highlights the import of prison space, through attention to the relationship between prison design and philosophies of punishment, in ways which resonate with, but differ significantly from, situational control research within psychology (e.g. Wortley 2002, Wener 2012). The recent collection *Architecture and Justice* (Simon et al. 2013: 1) takes a highly 'geographical' approach to spaces of justice, arranging its chapters in 'escalating increments of scale', 'from the prison cell ... to the social realm'. Essentially, there is increasingly a sense in which carceral space is being recognised and understood as critical to an engagement with the interior life of prisons.

If the interior spaces of prisons are increasingly being viewed by criminologists, prison sociologists and prison ethnographers as more than just containers for the 'experiences and practices that few other members of society have the opportunity to see' (Drake and Earle 2013: 12), then there is also an appreciation that the prison is not the only site with the 'social realm' under the influence of incarceration, or, as Smith (2013: 167) has put it, that the 'penal state is operative in sites where we might not be accustomed to look for it: not only within the prison interior ... but also, peculiarly, in cities that have been emptied of their 'troublesome poverty' and transformed into smooth, clean zones for the enjoyment of "consumers of urban space"'. Reviewing Wacquant's (2009) *Prisons of Poverty*, Smith (2013) was intrigued by his accounts of public space, especially the space of the metropolitan centre, and the ways in which the spaces of the prison open out into urban spaces of marginality in the context of hyperincarceration. These 'smooth, clean zones', of course, are the results of the exclusion of criminalised underclasses from the affluent 'forbidden cities' described by Davis (1990, 2006), in which the security infrastructure of the prison seeps into urban space in complex ways. As Shabazz (2009) has argued, security infrastructure, such as barred windows and turnstiles, installed in public housing, vividly recalls carceral spaces, and thus acclimatises young men to imprisonment, with 'hyperpolicing' converting impoverished inner-city areas into intensely regulated, 'prisonlike' spaces. At the same time, the 'forbidden' cities of affluent neighbourhoods are also protected by security technologies, in this context generating insulated spaces 'rich with atmospheres of wellbeing' (Adey et al. 2013: 301).

Hyperincarceration and the 'New Punitiveness'

It is perhaps no surprise that the emergence of carceral geography as a vibrant subdiscipline of human geography coincides with major changes in the scale and the scope of incarceration, in many parts of the English-speaking world, with regard both to those imprisoned by the prevailing legal system, and those detained in relation to their migration status. In what Wacquant (2011a: 3) has described as 'the great penal leap backward', the United States has the dubious position of leading the world in towards mass incarceration, or hyperincarceration, with the largest prison population in the world (Bonds 2012: 129) at 2.2 million[2] inmates at the time of writing, and a prison population that grew sevenfold between 1970 and 2003. Table 2.1 reproduces published incarceration rates for the carceral contexts discussed in this volume, demonstrating the US' position in global context.

The impacts of mass incarceration, and in the US in particular its highly racialized incarcerative practices, are felt far beyond prison walls. Wacquant's (2011b) thesis is that the penalization of poverty seen in the US in recent decades (and arguably extending into Western Europe and elsewhere through the exporting of US penal ideas and management systems [Downes 2007: 118, in Gottschalk 2009]) comprises a 'punitive revamping' of public policy by tackling urban marginality through punitive containment. Hyperincarceration, having in the United States thrown its 'carceral mesh' (Wacquant 2011b: 13) around the hyper-ghetto, is argued to have established a 'single carceral continuum' between the ghetto and the prison system in a 'self-perpetuating cycle of social and legal marginality with devastating personal and social consequences' (Wacquant 2000: 384).

---

2   Walmsley, R. 2013. *World Prison Population List*, 10th Edition (data from US Bureau of Justice Statistics 31/12/11).

Table 2.1   Incarceration rates, 2012

Country	Prison population total (number in penal institutions including pre-trial detainees)	Prison population rate (per 100,000 of national population)
USA	2,239,751	716
Russian Federation	681,600	475
South Africa	156,370	294
Uruguay	9,524	281
Colombia	118,201	245
United Kingdom (England and Wales)	84,430	148
Australia	29,383	130
China*	1,640,000	121
Canada	39,132	117
Ireland	4,495	100
United Kingdom (Northern Ireland)	1,851	101
France	62,443	98
Germany	64,379	79
Norway	3,649	72
Sweden	6,364	67
Finland	3,134	58
Iceland	152	47
Nigeria	54,144	32

Note: * Sentenced prisoners only. The total does not include those in pre-trial detention or 'administrative detention'.
Source: Walmsley, R. 2013. *World Prison Population List*, 10th Edition. International Centre for Prison Studies, University of Essex.

The development of carceral geography's concern with the 'new punitiveness' (Pratt et al. 2011), often taken to refer directly to more austere and overcrowded prison conditions, longer sentences, increased criminal sanctions and 'humiliating' punishments, has been paralleled within human geography more broadly by observations of the impact of increasingly punitive policies, for example, towards the treatment of the homeless (De Verteuil et al. 2009, Laurenson and Collins 2007), urban crime (Herbert and Brown 2006), zero-tolerance policing (Swanson 2013) and sexual deviance (Craddock 2000). Much of this work appears under the heading of what Sparke (2006: 357) has called 'the big 'N' of Neoliberalism', an umbrella term for the 'diverse ideologies, policies and practices associated with liberalizing global markets and expanding entrepreneurial practices and capitalist power relations into whole new areas of social, political and biophysical life'. Sparke's (2006) contention, after Barnett (2005), was that the use of the 'codeword' neoliberalism arguably masks and exacerbates the challenges of contemporary political engagement, and he urged geographers to nuance analyses of neoliberalism through analysis of the context-contingent connections between neoliberal governance and neoliberal governmentality.

The emergence of carceral geography at this specific juncture, in relation to the punitive turn and trends towards (and also away from) mass incarceration seems to be closely aligned to Sparke's (2006) steer. Drawing direct connections between neoliberal governance and the organised practices and techniques through which subjects are governed via the criminal justice system and the carceral estate, carceral geography has the opportunity to contribute significantly to understandings of governance and the geographies of control.

Structure

This outline of carceral geography's emergence and dialogues is intentionally brief. The sections of the book which follow both expand upon the discussions initiated above, in terms of providing more discursive accounts of scholarship in carceral geography, and suggesting future themes which could shape research in this field.

PART I
Carceral Space

Chapter 3
Carceral Space

This chapter provides an overview of existing geographical scholarship on the experience of carceral spaces, in order to foreground the following chapters which suggest future directions in which this scholarship may move.

There is a long-standing recognition that the nature of prison spaces enables actual or perceived constant surveillance, and that this situation has a direct influence on prisoner behaviour and control (Foucault 1979, Alford 2000). However, in the literature on the built environment of prisons, this issue has until relatively recently (e.g. Jewkes and Johnston 2007, Fiddler 2010, Hancock and Jewkes 2011), remained under-researched (Marshall 2000, Fairweather and McConville 2000). Human geographers studying carceral spaces bring to this context a specific understanding of space, recognising it as more than the surface upon which social practices take place (Gregory and Urry 1985, Lefebvre 1991, Massey 1994); they understand it instead as simultaneously the medium and the outcome not only of political or macro-economic practices, but also of everyday social relations across all spatial scales (Soja 1985). As Adey (2008: 440) argued 'specific spatial structures ... can work to organise affect to have certain effects upon motion and emotion'. Designers of spaces consider what Rose et al. (2010: 347) called 'seductive spatiality', or 'ambient power' (after Allen 2006: 445) which serve to shape behaviour in these spaces (Allen 2006, Adey 2008, Thrift 2004). In a carceral context in which spaces of imprisonment might commonly be assumed, in the lexicon of Foucault and Agamben, to deliver 'docile' bodies and 'bare life', geographers engage with the notion of agency within carceral space, identifying ways in which those confined within these spaces may resist regimes of incarceration and deploy spatial strategies to access and express agency.

The work of Michel Foucault and Giorgio Agamben undoubtedly underpins much recent human geographical scholarship engaging with spaces of confinement, particularly in terms of understandings of the nature of agency within carceral spaces. Whilst there is not scope within this chapter to reproduce a fine-grained exegesis of these theoretical constructs and the relationship between them which has developed within contemporary human geography (e.g. Crampton and Elden 2007, Minca 2006, Reid-Henry 2007), the chapter focuses instead on the ways in which their work has informed, and has been critiqued in, geographical scholarship of carceral settings. It unfolds as follows; initially, by outlining in broad brush the work of Foucault and Agamben as it pertains to understandings of carceral spaces and agentic practice within them, and next by discussing the ways in which these understandings and critiques have shaped the contemporary focus of scholarship of carceral spaces undertaken by human geographers.

Biopolitics, Sovereign Power, Space and Agency

Much scholarship on carceral spaces and agency within them flows from the works of Michel Foucault and Giorgio Agamben, or rather, from interpretations of their theorisations of biopolitics, as they relate to space, and to the relationship between space and agency. This section of the chapter outlines first the works of Foucault and Agamben most commonly used in this way, and then the ways in which these have been taken up and worked through by human geographers researching carceral space.

In the much-cited *Discipline and Punish*, Foucault traced the seventeenth- and eighteenth-century emergence of the prison as a new relationship between punishment and the human body; in contrast with the spectacle of physical torture and execution which was an 'art of unbearable sensations', the modern prison represented 'an economy of suspended rights' (1979: 11). For Foucault, the purpose of the reform movement which led to the development of the prison was not so much to devise a proportioned or equitable system of punishment, as to establish an economy of the power to punish, minimising political and economic cost. This new 'penal semiotics' established a relationship between clarified legal procedures, determined sentences, a rational legal process and critically, the individualisation of punishment. Foucault argued that the seventeenth- and eighteenth-century reformatory prisons, which operated a system of control of prisoners through organised schedules of work, premised on a belief in the possibility of rehabilitation of individuals, paved the way for a penal system which manipulated both the body and the soul, seeking to deliver an obedient object. He used the term 'docile bodies' to convey the object of power in these penal settings; under a minute and subtle control based on uninterrupted and constant observation in a 'codification that partitions as closely as possible time, space, movement' (ibid. 137).

Most famously, Foucault described Bentham's eighteenth-century *Panopticon* design for a model prison as a complete panoply of the disciplinary techniques which delivered these 'docile bodies', and used the term 'panopticism' to label the discourse which emerges from it and broadens into an understanding of the 'disciplinary society'. The Panopticon, with its central observation tower from which the peripheral cells in a circular building could be viewed, enabled constant surveillance of inmates by an unseen observer, without direct awareness of being watched. Foucault argued that the Panopticon was 'a machine for creating and sustaining a power relation independent of the person who exercises it' (ibid. 201), in that inmates internalised the regime; through regulation of space, segregation of individuals, and unseen but constant surveillance of the body, the subject is moulded into its own primary disciplinary force.

Although Foucault's account of biopower bears close resemblance to Giorgio Agamben's, the relationship between them is complex, and indeed is considered by some to be irreconcilable (Ojakangas 2005). The cleavages between them are, as Snoek (2010) has argued, based on understandings of the relationship between biopolitics and sovereignty. Whereas in *Discipline and Punish*, Foucault described the *displacement* of sovereign power by disciplinary power, Agamben interprets biopower as *closely related* to sovereignty. Rather than sovereign power giving way chronologically to biopower, as Foucault initially argued (before later modifying his position to one in which they exist in a complex interrelationship), Agamben saw the production of the biopolitical body as 'the original act of sovereign power' (Snoek 2010: 48). In *Homo Sacer: Sovereign Power and Bare Life* (1998) he posited that contemporary sovereign power deploys biopolitics to reduce citizens to 'bare life' and to hold them in a generalised 'state of exception'. His geneaology of sovereign power differs from Foucault's in that he traced its 'historical shadow' (Rieger 2005) back to the Roman Empire, with 'bare life' originating as *homo sacer*, a banned being who under Roman law was not eligible for ritual sacrifice, and could be killed by anyone with impunity (Agamben 1998). Bare life was thus defined by a dual exclusion from both political and sacred space, a life reduced to mere biological existence, stripped of all political status and held in a 'state of exception' where the rule of law is permanently suspended (ibid. 96). Agamben argued that such a space, which at once excludes bare life from, and captures it within, the political order, is the hidden foundation of sovereign power. He further argued that spaces of exception in which bare life prevails have existed since the time of *homo sacer*, and continue to exist today. To explicate the state of exception, he used the notion of 'the camp' – 'as the pure, absolute, and impassable biopolitical space (insofar as it is founded solely on the state of exception) – [which] will appear as the hidden paradigm of the political space of modernity, whose metamorphoses and disguises we will have to learn to recognise' (Agamben 1998: 73). In other words, bare life and spaces of exception exist in multiple concealed forms within the political space in which we now live, taking form in, for example, Guantanamo prison, and detention centres for illegal migrants (Agamben 2005).

Foucault and Agamben may have diverged over the precise relationship between biopolitics and sovereign power, but critiques of their approaches share some ground. Although, as Driver (1985: 426) observed, Foucault saw spatial organisation as 'an important part of social, economic and political strategies in particular contexts', in that 'the control and division of space become a vital means for the discipline and surveillance of individuals', his understanding of the prison itself was based on very little direct experience inside carceral institutions. Given the immense influence of the extension of his theory of biopower into the social body via the 'carceral archipelago', it is perhaps unsurprising that since the publication of *Discipline and Punish*, renewed attention has been paid to the experience of carceral space, with researchers 'testing' Foucauldian assertions of the production of disciplined or docile bodies in carceral settings, and in many cases finding them wanting.

Similarly, although Agamben's work is considered to be particularly pertinent post-9/11, critics have questioned his monolithic vision of the modern spaces of bare life, and have highlighted a perceived lack of proportion when considering under one heading a variety of spaces including refugee camps, concentration camps and gated communities (Levy 2010). Most importantly for engagement with experience of carceral spaces, it is also argued that Agamben fails to recognise the complexity of life within 'the camp', and in particular elides acts of resistance and other practices which operate to maintain humanity and resist bare life within the state of exception (Bigo 2007, Walters 2008).

Within human geography, then, initial explorations of spaces of imprisonment which engage directly with Foucauldian approaches have problematized the extent to which the notion of the prison as a constantly surveilled space in which prisoners internalise the regime to become 'docile bodies', actually represents contemporary prisoner experience. Human geographers engaging with Agambenian constructs similarly question the extent to which carceral space can be understood as a biopolitical space where sovereignty creates 'bare life'. In both cases, critiques emerge which draw attention to the operation of agency in carceral space, and which challenge either the construction of 'bare life' after Agamben, or of 'docile bodies' after Foucault.

In so doing, it is arguably the case that geographers have read Foucault too literally, or too narrowly. As Huxley (2006: 192) has observed, he did not intend his discussion of Bentham's Panopticon to be read as a literal description of an actual prison:

> If I had wanted to describe "real life" in the prisons, I wouldn't indeed have gone to Bentham. But the fact that this real life isn't the same thing as the theoreticians' schema doesn't entail that these schemas are therefore utopian, imaginary etc. One could only think that if one had a very impoverished notion of the real ... It is absolutely correct that the actual functioning of the prisons ... was a witches brew compared to the Benthamite machine. (Foucault 1991: 81, cited in Huxley 2006: 192)

Nevertheless, the Panopticon is 'a diagram of a mechanism of power reduced to its ideal form ... it is in fact a figure of a political technology' (Foucault 1979: 205). Such a model serves, as Elden (2001) has argued, as an aspiration against which achievement may be measured, and also as a distillation of the underlying logic of multiple and dispersed practices (Huxley 2006: 194). Whilst the prison as it actually exists could arguably never be expected to replicate perfectly the conditions of the Panopticon as envisaged by Bentham, geographers who have explored actually-existing prison spaces have tended to do so with the *intentions* of the Panopticon in mind, and in many cases have structured their work to consider the extent to which the prison under study *aspires* to the 'Benthamite machine'.

'Docile Bodies'?

Teresa Dirsuweit's (1999) paper on the experience of women in a South African prison was perhaps the first within human geography to directly engage empirically with Foucauldian perspectives of the spatialized mechanisms of power and discipline. She first highlighted the apparent similarities between the prison under study and the model Panopticon, describing it as 'subject to the constant surveillant control of light' (1999: 73), with surveillance in the prison organised both architecturally, via a central graded *spiraal* in which 'any position in the *spiraal* affords the viewer a line of sight which includes most of the *spiraal*' and ensures that 'any activity in the *spiraal* is observable from a vantage point which may not be observed', and by a lack of physical boundaries in living spaces; and socially, through a system of prisoner spies or 'pimps' who operated on the accommodation 'sections' which radiated out from the *spiraal*. Dirsuweit argued that the normative function of the prison was directly linked to its spatial organisation, in that the surveillance facilitated by the *spiraal*, allied with the lack of sensory stimulation in a drab and austere interior, and a lack of privacy in living accommodation, meant that the identities with which prisoners entered the prison became eroded over time, to be reshaped by context-appropriate 'feminine' rehabilitative pastimes such as sewing and cooking.

In this way, she argued, the prison fulfilled the 'omni-disciplinary' function of the complete and austere institution to normalise the delinquent (ibid. 75) after Foucault. However, the majority of her paper is taken up with vivid depiction of the ways in which prisoners did *not* conform to this 'ideal of rehabilitation and normalisation'; the ways in which they actively *resisted* the 'continuous and repetitive remoulding of their identity', and subverted the constant supervision of those in authority. The *spiraal*, emblematic of the classic Panopticon, is described as the central site of prisoners' 'flagrant disregard' for its unseen surveillance, and as the hub of a vibrant prison culture that defied efforts to 'alienate, confine and discipline' (ibid.). She argued that prisoners had 'appropriated the meaning of this space from a place intended to discipline their behaviour ... to a public arena for themselves through everyday activities' (ibid. 76) and were willing to defend their perceived 'right' to occupy and use this space by rioting when access to it was threatened by the prison authorities. In their living accommodation, prisoners created private spaces by curtaining off their beds with bedsheets and signifying to the 'pimps' that a 'home' had been created out of their view. Thus the Foucauldian notion of 'docile bodies' engendered through constant surveillance and internalisation of the regime seems rather far removed from the actual day-to-day activities of this prison and its inmates.

Ten years later, and distancing themselves somewhat from Foucauldian perspectives, in the preface to discussion of their study of boundary construction and personal space in a US prison, Sibley and van Hoven (2009) reviewed critiques of Foucault's interpretation of prison space. They pointed to Vidler's (1993: 84) observation that following Foucault, attention had been directed to the 'transparent space theorised as a paradigm of total control', understood as 'hygeinic space' by modernists after Le Corbusier, and noted that by contrast, their own work suggested the existence of 'material and imaginary spaces, that are unseen and not susceptible to regulation by the regime' (Sibley and van Hoven 2009: 199). They also highlighted critiques of Foucauldian perspectives, notably the operation of self-surveillance, rooted in the interpretation of the panoptic tower and the knowledge of constant unseen observation. One interpretation of the tower, they argued after Alford (2000: 129) was that it functioned as a 'carceral superego', 'in the sense that it required the prisoner to engage in self-surveillance without thinking' (Sibley and van Hoven 2009: 199), in that the disciplinary urge becomes part of the individual prisoner's unconscious, changing who they are. Referring to criminological literatures on prisoner socialisation, they pointed out that although the extent to which this process leads to prisoner 'docility' must depend on a prisoner's personality, the length of time they spend in prison, and their social connections within and outside the prison, this perspective on surveillance has been overlooked. Connectedly, they argued that seeing the prisoner not as an 'isolated and insulated product of subjection, as in Bentham's idealised Panopticon', but instead as an active agent, raised further concerns with the Foucauldian perspective. Referring to the work of Vaz and Bruno (2003), they argued that prisoners may *seem* to resemble 'docile bodies', but that this docility may be simply a mask that prisoners wore as long as they *thought they might be* under observation, or in other words, that prisoners *feign* conformity with the regime but do not actually internalise its values.

Sibley and van Hoven's work pointed to two key areas of critique of Foucauldian perspectives. First, the agency of prisoners as rational actors and social beings; they argued (van Hoven and Sibley 2008: 1004) that Foucault's interpretation of the Panopticon essentially elides the subject of surveillance as a being with agency. Second, the importance of the geography of the prison; here they referred to Alford's (2000) critique of Foucault, based on extensive fieldwork in a Maryland prison in the United States. Alford concluded that the Foucauldian thesis of docile bodies produced through surveillance and self-surveillance simply did not apply in the maximum security Patuxent prison in which he worked. He argued that in fact, the opposite was the case – 'not only are the disciplinary practices absent, but what is, in effect, the opposite principle reigns: if you control the entrances and the exits, you do not have to look' (Alford 2000: 127). In Patuxent, death row or long sentence prisoners could not escape, and minute surveillance therefore had no purpose. Whereas Alford (2000) hinted that the internal geographies of the prison are of little import, Sibley and van Hoven (2009: 200) concluded that the 'distinctive geographies of the institution need to be recognised in attempts to understand socio-spatial relations'. The point here is that the *nature* of the prison spaces and their *purposes* do matter, and they matter intensely.

These two imperatives – to recognise prisoners as agentic beings, and to focus on the geography of carceral spaces – have shaped much of the recent scholarship within contemporary carceral geography. Sibley and van Hoven's own empirical work combined these two intentions, drawing intriguing conclusions about the relationship between the internal geography of the institution and the social relations which exist within it. They focussed on the significance of the scopic regime, distinguishing between 'surveillance' (as the centralised institutionalised form of control) and 'looking' (as a diffuse and informal aspect of people's relations with each other). They argued that in day-to-day coping and carving out of spaces for themselves in prison, people make frequent visual assessments of others 'as part of a process of avoiding or associating with other inmates, in an environment of uncertain and sometimes volatile interpersonal relations, as well as thinking about being watched during the daily routines of prison life' (van Hoven and Sibley 2009: 1001–2). Vision, they argued, is of central importance for an inmate's ability to 'make space' within a carceral environment. The ways in which inmates look, are seen, or stay out of sight are in turn affected by the materiality of the carceral environment, and as van Hoven and Sibley (2009) pointed out, these vary enormously both within and between carceral estates.

The visual was also key to Baer's (2005) work on the personalisation of prison space in the UK. Noticing the accumulation and display of personal items such as toiletries (including empty toiletry containers), air fresheners and magazines within prison cells, he considered this practice in the light of de Certeau's (1984) theorisation of tactics, debating whether the stockpiling and display of these items could be interpreted as a sign of ownership and wealth within the informal economy of the prison, as a status symbol of a prisoner's personal hygiene; simply something to disguise the ugliness of the surroundings, or very practically, in the case of air fresheners, to mask the odour of drugs being smoked in cells. He noted that at interview, prisoners had described giving their collections of empty containers to friends, who then displayed them within their own cells, and he considered whether these accessories then became some kind of 'relic' which could 'conceivably hold memories from old friends and experiences' (2005: 213). Baer noted that what he was observing was 'a *visual imprint* of tactics on the landscape of the prison' (2005: 214) (emphasis in original). Although not directly positioned in relation to Foucauldian perspectives, by drawing attention to prisoners as individuals using tactics to personalise carceral spaces, his work implicitly highlights both their agentic practice and its close association with their specific material environment.

Spaces of Exception?

Whereas Foucault discussed the prison, as a metaphor for society at large, Agamben's theorisations are based on the 'camp', which he saw as fundamentally different from the prison. Until relatively recently, the nature of Agamben's theorisation of the camp, as a space outside of the juridical order in which bare life is rendered devoid of all vestiges of citizenship, has meant that its application to *prisons* has been problematic. Agamben himself specified that 'the camp – and not the prison' is the space that corresponds to the state of exception, since 'while prison law only constitutes a particular sphere of penal law and is not outside the normal order, the juridical constellation that guides the camp is ... martial law and state of siege' (1998: 19). The camp is 'topologically different from a simple space of confinement' (ibid.). Although confined, in theory, prisoners maintain certain aspects of citizenship.

However, recent developments in imprisonment in the United States, as Czajka (2005) has contended, and arguably the dissemination of these practices more broadly, mean that the temptation to conflate the prison and the camp becomes ever more justified. She argued that with the rise of supermaximum confinement ('supermax'), an environment now exists in the United States where two categories of prison can be identified. 'Standard' prisons, (characterised by prison labour, a sense of rehabilitative rhetoric, and retention of some aspects of citizenship despite the partial removal of citizenship rights), represent institutions in which prisoners remain 'within the purview of legal and social structures that restrict the realm of possibility with respect to their treatment and living conditions' (ibid. 130). These correspond to the 'prison' as envisaged by Agamben – controlled by prison law as a subset of penal law, and existing within the normal order. Although stripped of most rights of citizenship, prisoners in this context do enjoy some protection via legal and social structures and are 'included through their exclusion in both the economic and political spheres of the society that expels them' (ibid.). In 'supermax' prisons, by contrast, she argued that prisoners exist entirely outside of the rubric of citizenship, deprived of all work, all education, all physical and mental stimuli, and individually confined to a cell with poured concrete fixtures. Rather than restricting opportunity and choice in a way which requires the definition of prisoners' rights and entitlements (such as the availability of TV stations, the ability to purchase food, and permission to decorate cells) supermax prisons simply remove everything whose presence could require any kind of choice to be exercised. As the supermax system has evolved, she argued, the exception has become the norm – the kind of 'lockdown' measures traditionally only employed temporarily and in exceptional situations within 'standard' prisons, have become entrenched and normalised within the supermax institution. She conceded that it is difficult to establish the extent to which these institutions genuinely operate outside of the purview of legal scrutiny; however she also pointed out that a range of decisions pertaining to supermax confinement are now taken outside of judicial oversight and without the right of appeal, and that statements from inmates suggest that they perceive themselves to exist at the mercy of correctional officers, and without legal protection. In this way, supermax prisons serve to delete prisoners from the sphere of citizenship altogether, and Czajka (2005: 131) argued, to recall the conditions of the camp after Agamben, in which residents of these spaces are so completely deprived of the rights of citizenship that 'no act committed against them appears a crime, everything is truly possible, and the state of exception is normalised'.

Agency, Governmentality, Subjection and Subversion

Most recently within carceral geography, increasingly complex and nuanced readings of Foucault and Agamben have been coupled with considerations of other theoretical approaches to confinement which emphasise the significance and complexity of agency within carceral space, and in bringing these into dialogue, have opened a space for meaningful engagements with the lived experience of incarceration.

In her work on asylum seeker hunger strikes, Conlon (2013) drew attention to Foucault's theorisation of the development and arrangement of power relations in society towards governmentality, in which she noted that even though governmentality operates to shape individuals' actions, one of its central features is the individual's capacity to exercise freedom, as a series of autonomous, though orchestrated choices.

The conceptualisation of autonomy is itself a point of tension. Many observers argue that it seems to be absent from Foucault's theorisations (Moran et al. 2012); the subject seemingly rendered impotent in the face of immanent and all-pervasive power, since it only 'comes into being as a construct of a regime of power' (Bevir 1999: 66). Wisnewski (2000) critiques this interpretation, arguing that '[t]he fact that we are always in relationships of power does not entail that we cannot make choices freely ... Existing in the exercises of power does not mean existing without choice – it means only that choices occur within a realm of possibilities defined by discursive practices' (Wisnewski 2000: 434). As Conlon (2013) noted, Foucault allows for a '"weak" autonomy, understood as the ability to make more or less free choices within a realm of possibilities' (Wisnewski 2000: 435), a situation described by Bevir (1999) as a distinction (elided, he contended, by Foucault) between autonomy and agency. For Bevir, Foucault's rejection of autonomy does not necessarily entail the rejection of agency. Since different people respond differently to the same social structures, there must, he argued, be an 'undecided space in front of these structures where individuals decide what beliefs to hold and what actions to perform' (1999: 68) effectively making choices within Wisnewski's realm of possibilities. The subject is, therefore, an agent, even if not an entirely autonomous one (Bevir 1999).

Whilst geographers have drawn heavily on Foucault's theorisations of biopower and the Panopticon, Conlon (2013) argued that there has to date been limited engagement with his writings on counter-conducts and critical attitudes in the context of incarceration. Her work draws attention to readings of Foucault which emphasise the central role of questioning the nature of governing as part of the practice of governmentality itself. Foucault calls such questioning 'counter-conducts' or a 'critical attitude', whose objective is 'a different form of conduct, that is to say: wanting to be conducted differently, by other leaders [...] towards other objectives [...] and through other procedures and methods' (Foucault 2007: 194–5, in Conlon 2013). These counter-conducts and critical attitudes are seen by Cadman (2010: 540) to be 'wholly immanent and necessary to the formation of governmentality', in that, as Conlon (2013: 141) wrote, 'questioning, critique and struggles around who governs, how and why individuals or populations should be governed in particular ways, and what technologies are used in this process are embedded in and crucial to how governmentality operates'. Further, because counter-conducts and critical attitudes are integral to governmentality, they do not entail rejection or negation of it. Rather they are an effort to open up – without any certainty as to the outcome – possibilities to govern and be governed differently.

In her own study of threatened hunger strikes among asylum seekers detained in Ireland, Conlon (2013: 144) understood this specific threat as a form of counter-conduct that raised questions about how the detainees were governed, and in so doing queried the effectiveness of existing technologies of government of asylum seekers in that context. It also delivered a concrete outcome in that the Irish authorities conceded to the detainees on the issue of movement which had initially prompted the threatened hunger strike. Conlon argued that asylum seekers' hunger strikes entail counter-conducts or critical attitudes which involve autonomy, subjugation, critique and desubjugation, each of which is immanent within governmentality; and that through these practices, they 'claim the right to question how they are governed and thus constitute themselves not merely as governed individuals but also as political subjects' (2013: 145). In making this argument, she contended that questioning who governs, why, and how, are not 'discrete, singular acts of agency' but rather 'contingent and continuous political practices that are embedded within the rationality and technologies of government' (ibid.). Although they may deliver a sense of empowerment, these counter-conducts are imbued with risk and uncertainty; unlike 'resistance' which is often viewed as progressive, or proceeding towards some improvement in a situation, counter-conducts cannot prescribe ahead of time what the results of questioning and critique will be. As Conlon (2013: 142) noted, 'it is on account of their immanence in governmentality [...] that counter-conducts and critical attitudes must be understood as ongoing political practices rather than as discrete or monumental acts of individual agency'.

This perspective encourages carceral geographers to pay ever closer attention to the nature of agentic practice within carceral settings. Whereas previously, interpretation of the activity of those confined within carceral space has tended towards identifying 'resistance' to becoming Foucauldian 'docile bodies', Conlon's work suggests that rather than seeing individual acts of agency which negate the categorisation of 'docile bodies', in terms of counter-conduct and critical attitude as part and parcel of governmentality, those bodies could be expected to exercise autonomy and still be constructed as 'docile'.

This counter-conduct or critical attitude bears close resemblance to what in *Asylums* (1961) Erving Goffman termed 'secondary adjustments' – a set of improvised means, and use by prisoners of resources such as material objects, times and places, to preserve a certain degree of autonomy, often in a clandestine manner. As for Foucault, who saw counter conduct and critical attitude as immanent within and critical to governmentality, in Goffman's theorisation, the 'total institution' and its insubordinate underlife were intrinsically bound together. The subjection of inmates to the control of the total institution, and their subversion of it, being two sides of the same coin.

This duality of subjection and subversion was at the heart of de Dardel's (2013) analysis of prisoner culture within the Colombian prison system, in which she critiqued the applicability of Agamben's notion of 'bare life' within the carceral setting, using Goffman's terminology of 'secondary adjustments' to describe the agency exhibited by prisoners in this context. For de Dardel, the consideration of Agamben is a crucial one – she argued that scholars of carceral spaces have paid too little attention to the relevance of his theorisations to these spaces, and that for Colombia, where US penal ideals are being imported as a form of 'best practice' policy transfer, there are genuine possibilities of the kind of normalisation of the exceptional described by Czajka (2005) in the United States. De Dardel's study describes the introduction to Colombia of the 'New Prison Culture', using techniques commonly practiced in the US, and bringing a new form of management of spaces and bodies, characterised by close control and severe restrictions, and its 'collision' (2013: 187) with the pre-existing *criolla* prison culture, in which the prison courtyard operated as the space in which the 'collective and self-organised life of the prisoners takes place, with very little presence of the prison officers' (ibid. 186). Noting that prisoners in her study actively resisted the imposition of the new rules, de Dardel drew particular attention to the embodied strategies which they deployed, as a direct counterpoint to Agamben's theorisation of inmates of the 'camp' being reduced to 'bare life'. Her female respondents described deploying embodied and gendered tactics of personal hygiene (for example, sucking water from leaking pipes in order to be able to wash more frequently than permitted by prison authorities). Both men and women, held separately within the same facility, engaged in a common practice of spelling out visual messages to one another through the twirling and coiling of towels around arms and legs protruding from barred windows. Men in a US-style 'model facility' described the 'loosening' of the initial regime since the prison had opened in 2002, following 'low intensity' daily resistance and the 'open collective struggles' of prisoners (ibid. 194). In this facility, the 'regime' had given way on prisoners wearing uniforms, having regular intimate/conjugal visits, and using the prison courtyard for collective recreation. De Dardel argued that in Colombia's new prisons, the refusal of 'bare life' within the prison and the deployment of various 'tactics to clothe the body' (2013: 195) represent an element of the embedded *criolla* prison culture, creating a hybrid prison culture in which 'two contradictory carceral subcultures interpenetrate' (ibid. 196).

Conclusion: Agency, Tactics, Resistance, Counter-conduct

From a starting point of initial engagements with Foucault's notion of 'docile bodies', drawn out of the model 'Panopticon' prison institution, and bringing to carceral space a critical human geographical approach which draws upon the prevalent understandings of space and spatiality as multiplicitous and heterogeneous, both 'abstract and concrete, produced and producing, imagined and materialised, structured and lived, relational, relative and absolute' (Merriman et al. 2012), geographers have brought an increasingly diverse range of theorisations to bear on carceral space, doing so on the basis of data generation within diverse places of incarceration, precisely at a time when Wacquant (2002) drew attention to what he called the 'eclipse' of prison ethnography in the United States.

The result of this diverse scholarship is arguably that the theorisations, particularly of Foucault and later of Agamben, which initially dominated explorations of carceral spaces, have either receded in prominence within this literature, or been nuanced significantly through both closer and wider readings. More recently, geographers have begun to characterise what they variously call the autonomy, agency, tactics, subversion, or resistance of those confined within carceral space, thinking of this through the lexicon of de Certeau (1984), in which, as Jewkes (2013: 128) noted, the powerful construct and exercise their power, but the weak tactically create their own spaces within those places, 'making them temporarily their own as they occupy and move through them'.

This focus on the occupation of spaces and the transitory and transient nature of that occupation throws up a range of possible future directions for research in this field. Informed by understandings of space and spatiality as lived and experienced, and of the integral relationship between space and time (Massey 2005), carceral geographers could consider the embodied experience of imprisonment, and highlight the temporal aspect of imprisonment in conjunction with the spatial through a focus on prison TimeSpace. The following two chapters accordingly take up these two themes, exploring first the corporeal nature of incarceration, and then the experience of carceral TimeSpace.

Chapter 4
The Emotional and Embodied Geographies of Prison Life

This chapter considers the experience of carceral space at an intensely personal level, tracing the ways in which the individual spaces of the prison elicit and facilitate different emotional expression, the ways in which the experience of incarceration is inscribed corporeally upon the imprisoned body, and the embodied strategies deployed by occupants of carceral spaces. In so doing, it foregrounds the personal, emotional subject in ways which are articulated with the development of emotional geographies (Lorimer 2008), and with the 'situated studies of sensuous, corporeal, kinaesthetic experience and mundane circumstances of materiality, sociability, connection and association' which are growing popular within what Lorimer terms, after Cresswell (2006: 56) 'geographies of bodily movement' (ibid. 556). The affective geographies of imprisonment remain under researched, but the work discussed in this chapter demonstrates the critical importance of consideration of emotion (and affect) in close relation to individual lifeworlds, and with close attention to differences in subject position and subjection.

Although this chapter is entitled the 'emotional and embodied' geography of prison life, the inherently emotional nature of embodiment is readily acknowledged, in that, as Davidson and Milligan (2004: 523) have argued, 'we need to explore how we feel – as well as think – *through* "the body"' (emphasis in original). The spatially mediated nature of the articulation of emotion is also of central importance here, as emotional relations and interactions shape personal geographies, and place 'must be *felt* to make sense' (ibid. 524). In their review of emotional geographies moving 'out' from the body itself, Davidson and Milligan (2004: 526) consider institutional(ised) emotions of 'total' institutions including prisons, and argue that in these contexts, emotional expression, if not experience, is more or less 'governed' by rules. If emotion itself is defined as the culmination of 'affect as a field of pre-personal intensity, feeling as that intensity registered in sensing bodies, and emotion as a socio-cultural expression of that felt intensity' (McCormack 2008: 414, after Anderson and Smith 2001, Thrift 2004, Anderson 2006), then in the carceral context the 'socio-cultural' is of critical significance, in relation to the 'rules' which govern the expression of emotion.

Expressing emotion as 'felt intensity' in prison is a highly charged activity. As Crewe (2009, 2011) has noted, and as Sykes (1958) suggested, the *possibility* of violence and predation between prisoners is just as debilitating as the *actual* level of aggression, and it is 'this insidious sense of threat that means that most prisoners describe the atmosphere of most prisons as tense and enervating, regardless of whether they are personally confident of their safety' (Crewe et al. 2014: 58). Quite apart from the relations between prisoners, prison staff police prisoner behaviour, and prisoners are subject to situational and bureaucratic controls which make them circumspect about their daily interactions, under what Crewe et al. (2014: 59) term a kind of 'bureaucratic gaze' that judges their actions and attitudes. These everyday socio-cultural circumstances in the life of the prisoner constitute a significant barrier to emotional authenticity and expression. In this context, the purpose of this discussion is to consider the recent work within criminology and human geography which focuses first on what might be termed emotional geographies, and the expression of emotion in carceral space, and then on embodied geographies of incarceration, in order to draw attention to the significance of emotional and embodied geographies for furthering understandings of the diverse experience of carceral space.

The 'Emotional Geography' of the Prison

Very recently, carceral geographers and criminologists have come into dialogue over the notions of public and private space (Moran et al. 2013b, Milhaud and Moran 2013), front stage and back stage, and the 'emotional geography' (Crewe et al. 2014) of prison life. Writing within criminology, Crewe et al.'s (2014) use of the term 'emotional geography' is something of a departure from the lexicon of human geography, meaning, as it does in this context, 'the zones in which certain kinds of emotional feelings and displays are more or less acceptable' (ibid.). Despite the use of terminology such as 'sterile areas', and 'segregation unit' within prison sociology they argued that 'very little work has taken seriously the idea that prisons are complex and spatially differentiated emotional domains, beyond the binary metaphor of front or back stages' (ibid.). The front and back stages to which they refer derive from Goffman (1959) and Giddens' (1984) notions; 'frontstage' being the public aspect of identity presented in social engagement with others, and 'backstage' being the restoration of the interior sense of self where 'frontstage' performance is no longer demanded (Jewkes 2005). 'Frontstage' and 'backstage' are very familiar concepts within criminology, with scholars widely observing that inmates adopt façades while 'inside', that this 'front' is impossible to sustain indefinitely, and that the facility to 'be oneself' at some point is essential for prisoner wellbeing. As Crewe et al. (2014) note, there is also a certain appeal in the ideas of front and backstage, in part because prisoners themselves are frequently reported to use the term 'fronting' to describe a persona they adopt to present themselves to others.

However, mapping these terms onto specific spaces or spatial contexts has proven problematic, in that distinguishing between discrete areas where 'frontstage' identities are adopted and where the 'backstage' can be revealed, is very difficult. Whilst Goffman (1961) described the cell as a personal territory, in which prisoners could presumably relax into a backstage persona, the fact that prisoners commonly share sleeping accommodation means that, as Jewkes (2005) has argued, they have to maintain public or quasi-public façades even in this apparently private setting. One way to try to understand the geography of the prison as experienced by prisoners is to think of it in terms of public and private space (Milhaud and Moran 2013), and to consider the extent to which privacy maps onto the 'backstage' and public space requires the 'frontstage' personae described by Goffman.

In this context, defining what 'privacy' might mean is a challenge in itself. The public/private distinction in the social sciences is complex – it does not represent a single paired opposition, but a complex arrangement of binaries, specific to particular contexts and subject to different interpretations and understandings. Although 'public' and 'private' realms are considered to map neatly onto spaces in which life is lived, there is no intrinsic set of characteristics for either of these types of space. Drummond (2000: 2379) perhaps comes closest to a categorisation, summarising the Western academic debate around these terms as follows: 'private space is considered to be a domestic space where social reproduction occurs more or less free from outright control by outside forces such as the state ... and public space is "out there", belonging to the whole community, although regulated by prevailing social and legal norms'. This inside/outside dichotomy has been analysed and critiqued, with feminist scholars pointing out the patriarchal character of the association between masculinity and the public, and femininity and the private/domestic, in the light of the fact that the domestic can be the locus of oppression rather than freedom. Scholarship on public spaces explores their position in mediation between public and private, masculine and feminine; Jackson's (1998: 188) work, for example, on the shopping mall as a domesticated public space challenged the public/private binary and suggested that engagement with the public and the private must be 'sensitive to the socially differentiated nature of these highly contested spaces'. Mitchell (2003: 182) strongly underlined the 'exclusionary citizenship' promoted in contemporary public spaces. Geographers have actively disrupted the boundaries between public and private space, destabilising the assumed binary between them; Fenton (2005) identified citizenship in private space; Blomley (2005) argued that public and private categorisations of space are fluid, rather than mutually exclusive and exhaustive; Allen (2006) identified 'privatised public spaces'; in Berlin, Tyndall (2010) challenged the assumed 'publicness' of shopping malls in Sydney, Australia; Kumar and Makarova (2008) argued for the 'domestication' of public space, and Staeheli (2010: 72) reminded us of 'the importance of putatively private spaces of public deliberation and action'.

Within carceral space, confiscation of privacy has long been understood to be a 'functional prerequisite' of imprisonment (Schwartz 1972: 229), and one of Sykes' 'pains of imprisonment' (1958). Foucault (1979) argued that in the constantly surveilled Panopticon, prisoners are almost always either in one another's presence, or in the sight of penal authorities. This quotidian lack of privacy can include forced exposure (such as in communal showers); forced spectatorship (exposure to others' lack of privacy); and violation of collective privacy (through imposed and exposed intimacies) (Schwartz 1972). Much recent discussion of privacy within criminology and prison sociology elides prisoner privacy *per se*, considering instead the implicit lack of privacy characteristic of prison overcrowding, where single-cell or small unit accommodation is perceived as a means of reducing noise, constant activity and violations of personal space which can arguably lead to stress responses and to an increased likelihood of disorder and violence (Schaeffer et al. 1988, Grant and Memmott 2007, Sharkey 2010).

Previous work in French and Russian prisons (Milhaud and Moran 2013) has attempted to consider the extent to which specific spaces within prisons in these two different penal systems are experienced as 'public' or 'private', and by extension, whether these spaces require or facilitate the performance of 'front' and 'backstage' prisoner identities. The comparison of Russia and France enabled two different philosophies of confinement translated into material form, to be held in comparison with one another. Whereas the architecture of French prisons promotes the separation of inmates through cellular confinement (Demonchy 1998, 2000, 2004), deterring inmates' free association, in the Russian system, prisoners are commonly accommodated together in communal detachment blocks; in this context deprivation of liberty should not involve isolation from society (Oleinik 2003). Despite the differences in penal architecture, in both contexts, prisoners described a complete lack of privacy, with the 'backstage' suppressed in France by the shared intimacies of cellular confinement, and in Russia by the psychological pressure of large numbers of people in the detachment blocks. Although in both Russia and France, prisoners described explicitly spatial strategies which they deployed to find or create private spaces in which they could 'be themselves' – such as remaining in the shared cell whilst a cellmate enjoyed time in the yard in France, or purposefully breaching prison rules to receive the 'punishment' of time in the isolation 'cooler' in Russia – they also reported using nominally 'public' spaces to retreat into themselves.

The construction of privacy within public space effectively blurs the boundaries between the two, and challenges the idea that 'front' and 'backstage' identities require mutually exclusive spaces in which to take form. In both contexts, prisoners described spaces which they shared with many others, such as prison workshops, or the prison yard, as places where they could choose to be alone with their thoughts, and that sometimes the atmospheres and ambient conditions in those spaces – in Russia, for example, the insulating noise of the prison sewing factory – facilitated this retreat into the self.

If a binary distinction of private and public, or 'backstage' and 'frontstage' is disrupted and challenged in this way by the personal experience and negotiation of prison space, then as Crewe et al. (2014: 62) argued, 'it is bound to over-simplify the prison's emotional universe and its spatial differentiation'. In their exploration of the 'emotional geography' of a UK men's prison, they found that although some prisoners reported a conventional distinction between frontstage and backstage when they relaxed in their cells listening to music, others reported revealing their emotions in specific places which Crewe et al. (2014: 64) term 'emotion zones'. In these spaces, which corresponded neither to conventional characterisations of 'frontstage' or 'backstage' areas, they observed a broader emotional register than was possible elsewhere.

The nature of these spaces is particularly significant, since these 'intermediate zones', as Crewe et al. (2014: 64) described them, shared specific characteristics which facilitated the suspension of the 'normal' rules of prisoner society about the masking of emotions and the modification of identity. These zones all existed beyond the everyday spaces of prison cells and landings which characterise the UK prison estate. Entering them required the prisoner to interact with individuals, be they prison personnel, instructors, visitors or volunteers, who represented something of the 'outside' world.

In a similar way, Smoyer and Blankenship's (2013) work on the significance of food in US women's prisons revealed the contribution of prison food systems to the construction both of boundaries that distinguish the spaces of the prison from spaces and lives outside of its walls, and of a 'patchwork' of places within the prison that each had their own structure, meaning, and associated micro-geographies that operated to define women's individual prison experience. They identified an illicit flow of smuggled food between the cafeteria, the kitchen, and the housing units, deconstructing internal boundaries intended to control the consumption of food, and transforming the housing units into spaces of ingenious food preparation, pooling of resources, and shared consumption, triggering complex social relations and dynamics, resonant with Ugelvik's (2011) work on surreptitious food production in Norwegian prisons. For Smoyer and Blankenship (2013), the study of food within prison revealed not just these complex micro-geographies, but also a sense that women's relative powerlessness and food insecurity (in the official prison food regime) encouraged engagement in illicit hoarding, smuggling and trading practices that mirrored many of the women's drug habits outside of prison. In deconstructing and breaking down the internal food boundaries, they argued, women overate, under-ate, and obsessed about food, replacing an addiction to drugs with an addiction to food, manifest in unhealthy eating behaviours which had embodied consequences for their physical health. The emotional geography of this carceral space, in terms of its complex system of places of acquisition, smuggling and consumption, enmeshed in a network of social relations based on mutual reliance and trust, related directly to women's emotional and embodied experience of incarceration. Similarly, for Carney (2013), studying feeding practices in migrant detention in the US, the body became a site of intervention for a biopolitics of citizenship and governmentality, in which food deprivation was a powerful instrument of social control. Food 'trauma' in detention centres served to compound existing traumatic experiences linked to food security (which may have impelled migration and asylum-seeking in the first place), in a practice which could lead to further disordered eating for years and generations to come, as well as reproducing the conditions of abjectivity and reinforcing zones of confinement for non-citizens.

The close relationship between the spaces of food consumption and the emotional geographies of imprisonment described by Carney (2013), and Smoyer and Blankenship (2013) draws attention to the inherently embodied experience of carceral space. Whilst this previous section of this chapter has considered physical spaces of incarceration, the discussion which follows explores the embodied experience of incarceration, which is deeply connected to the experience of carceral space – the physical body being the means by which spaces are directly experienced. In so doing, it engages with two closely related concepts – the embodied *inscription*, and the embodied *strategies* of incarceration, and deals with these in turn.[1]

The Embodied Inscription of Incarceration

In her work on the ageing female body in prison, Wahidin (2002: 178) used a Foucauldian analysis to demonstrate how discourses act upon and inscribe the female body, with the body held in a carceral prism in which power relations have an immediate hold over those under the prison gaze. Considering the prison as a total institution, she demonstrated how time as a technique of discipline is used by the prison to mark the body, and how the corporeality of time and its use in prison 'transcends the dualisms between subject/object and mind/body' (ibid.). Wahidin's work is informed by a conceptualisation of the self in a post-modern society as an unfinished project, one that is central to a person's sense of self-identity, with the body 'inscribed by variables such as gender, age, sexual orientation and ethnicity, and by a series of inscriptions which are dependent on types of spaces and places' (Wahidin 2002: 180). She drew upon Turner (1995) in emphasising key processes that work on and in the body across time and space, and argued in line with Giddens (1981) that the corporeal existence of the body is complex and reducible neither to biology nor the dictates of capitalism, occupying instead a position between the biological and the social, the collective and the individual, structure and agency. Wahidin translated these discourses into the prison setting, showing how prison time is inscribed upon the confined body. In so doing she compellingly conceptualised the body not as a passive materiality that simply awaits inscription, but rather emphasised its negotiation of the 'capillaries of power, enabling the body to be always in the process of becoming through the experiences of embodiment' (Wahidin 2002: 181).

Wahidin (2002: 192) argued convincingly that the material body almost becomes a medium through which power operates and functions; but that rather than being a passive medium it is rather interwoven with and constitutive of

---

1   Drawing from Moran, D. 2012. Prisoner Reintegration and the Stigma of Prison Time Inscribed on the Body. *Punishment & Society* 14: 564–83, and Moran, D. 2014. Leaving behind the 'total institution'? Teeth, TransCarceral Spaces and (Re)Inscription of the Formerly Incarcerated Body. *Gender, Place and Culture* 21(1): 35–51.

systems of meaning, signification and representation. Time and experience are, of course, corporeally inscribed (Ahmed and Stacey 2001) whether a body is incarcerated or not, such as through stretch-marks, wrinkles, and surgical scars. However, there is a mutually constitutive relationship between bodies and spaces (Longhurst 2005) in which bodies are understood as sites of 'textual inscription' (Johnson 2008: 563) which shapes identities and social relations as well as the conceptual and actual spaces in which bodies move. Bodies become inscribed in a way which suits or is an adaptation to, a specific space and/or circumstance (such as scarification which 'fits' the ritual practices of specific groups), but which may not suit another. McDowell (1999: 61) noted that bodies therefore may not necessarily 'fit' the idealised representations for certain 'spaces'; considered ill, ugly, wrong or deviant in a variety of ways. Although bodily inscription *per se* is not prison-specific, in that it takes place whether or not a body is incarcerated, the argument is that incarceration has a particular set of prison-dependent, tell-tale inscriptions – and that the stigma prisoners may experience after release is to some extent enabled by the 'lack of fit' between these inscriptions and the circumstances of release. Although the actual inscriptions are prison-dependent rather than prison-specific, the fact that prisoners perceive them in this way enables some insight into the corporeality of incarceration.

One frequently observed bodily change during imprisonment is a deterioration in dental health. Research conducted in a variety of geographical contexts (for example in China [McGrath 2002], Australia [Osborn et al. 2003], the US [Cunningham et al. 1985, Salive et al. 1989, Mixson et al. 1990, Boyer and Nielson-Thompson 2002], Nigeria [Dhlakama et al. 2006], and South Africa [Naidoo 2004]) drew similar conclusions; that prisoners' oral health is worse than that of the relevant background population. This finding is commonly attributed both to some prisoners' already poor oral hygiene before imprisonment, and also to the often poor provision or uptake of prison dental services. Studies also commonly find a tendency for prison dental services to extract teeth rather than fill cavities, resulting in the unnecessary loss of teeth, and accordingly, to poor dental health amongst prisoners, which tends to worsen as sentences progress, and contributes to poor general health and poor nutrition. Dental health has implications beyond physiology; research into self-rated oral health and measures of general health, self-esteem and life satisfaction, broadly finds that self-rated oral health is a good predictor of self-esteem and general wellbeing, and vice versa, i.e. that the better a person feels about their life in general, the more likely they are to take care of their teeth (Macgregor and Balding 1991, Benyamin et al. 2004, Locker 2009). Setting out in broad terms the relationship between prisoners' oral health and self-esteem, Williams (2007: 84) pointed out that in the US, missing teeth 'is becoming a tell-tale sign of having been incarcerated', and that poor oral health constrains social, professional and personal relationships, and reduces the likelihood of employment after release.

In the Russian Federation, although very little is known in general about the oral health of released prisoners,[2] previous research into women's experience of incarceration (Moran 2012, 2014) has found that incarcerated women feel particularly keenly the changes to their dentition during imprisonment. If prison dental services are usually worse than those provided to the general population, then in Russia they are particularly poor. As Jargin (2009: 519) has noted, some dental practitioners have continued a Soviet-era tendency towards invasive treatments, some manipulate patients 'towards extractions and prosthetics (dentures)', and minimally invasive techniques which seek to conserve dental tissue have only recently been introduced.

Missing teeth are considered a conspicuous stigma of imprisonment. Women interviewed in Russia after release from prison spoke evocatively about the significance of their dental health both for their self-esteem, and for their prospects of finding work and therefore successfully adjusting to life on the 'outside'. Their comments, discussed below, describe compellingly the simultaneously very personal and overtly public nature of missing teeth as a marker of prison time; as an example of prison time inscribed on the body in a way that was very embarrassing and stigmatising, and very difficult to hide.

According to prisoners' recollections (elicited during a study detailed in Moran 2014), Russia is no exception to the general observation of poor and invasive dental treatment in prison. Although one former prisoner reported that where she had served her sentence, the prison dentist had treated both prisoners and staff alike, and had filled teeth as well as extracted them, most women's experience was that although a dentist visited the prison once or twice a month, prisoners' dental health was poor and dental treatment was limited to extraction.

There were dentists, but dentists only extracted teeth.

Q. What do you mean?

A. Just that. We did not get fillings …

Q. So, if for example …

A. If there was toothache, then you'd go to the doctor. She would take a look and say – if you want to put up with an extraction, I'll pull it out. So she pulled it out.

Q. So the women were practically without dental care?

A. Yes.

---

2   The only indication comes from a 1991 study carried out in Magadan, in the Russian Far East, in which Hardwick et al. (1993) noted that the presence of a significant number of recently released prisoners in their sample of the local population impacted on their data.

Q. Why do you lose your teeth [in prison]?

A. Why do you lose them? If you get caries [tooth decay] they pull them out, just pull them out, but they will not drill them.

Q. So there isn't any treatment?

A. No.

The dentist just pulls teeth out. Actually they never filled them, just pulled them out. That's all [the dental treatment] that we had.

Q. Well, at least they are extracted with anaesthesia?

A. Yes, yes, of course.

In some prisons, dentists would fill cavities, rather than extract teeth, but only if prisoners could pay, out of the meagre wages they received for prison work.

For your teeth, if you have money you can go to the "hospital" [the prison clinic] for treatment. Otherwise they just pull out the aching tooth. They drill, they drill there. You can get the drill in the "hospital", for money.

The outcome of all of this is that women found themselves re-entering society with missing teeth and poor oral health, something that they collectively recognised as a problem. As one woman put it,

Teeth, teeth, teeth, teeth … There was virtually no one who didn't complain about the state of their teeth. Yes, they'd fall out, yes, they crumbled away. I lost two teeth while I was inside. It was the bad environment, the lack of vitamins, well everything, generally.

Missing teeth are just one of a range of physical characteristics and habits which are considered to 'mark out' ex-inmates, including tattoos, smoking and swearing. However, they seem to be the most problematic, in terms of the problems women face in concealing these bodily inscriptions of incarceration. Smoking, drinking and swearing are habits which can be difficult to break, but women can at least conceal them in public if they feel that this will help hide a personal history about which they feel self-conscious. Tattoos indelibly mark the skin, but are not necessarily publicly visible. However, any social interaction which involves conversing with others, and any public-facing employment, reveals missing teeth.

The Embodied Strategies of Incarceration

Although the intrinsically embodied nature of imprisonment is not perhaps immediately apparent, embodied practices or strategies, such as the 'dirty protest', in which faeces are smeared on cell walls, and hunger strikes, in which the frailty of the body is instrumentally exposed as a means to draw attention to perceived injustices, are amongst the most prominent expressions of resistance to carceral regimes. These examples of embodied protest and resistance have received considerable scholarly attention (such as Conlon 2013, Aretxaga 1995, Yuill 2007) but away from these 'extremes', there is scope for a focus on what Connell (2012: 867) has called the 'multiple narratives of embodiment', that is, the diverse personal corporeal experiences of imprisonment and the ways in which various quotidian embodied practices shape experiences of spaces of incarceration, drawing upon research from a range of carceral contexts.

This section of the chapter considers the work of Caputo-Levine (2013: 169) who described what she terms the 'carceral habitus', a situational adjustment to prison life that includes the adoption of embodied strategies such as a specific language, and changes in the relationship to the world and understanding of the body, as a structuring logic for understanding embodied strategies. Following Bourdieu, she discussed the understanding of the prison as a field, a set of power relations, or 'socially structured space' (Bourdieu and Wacquant 1992: 17) which presents a structure of probabilities and necessities, generating objective conditions which are then embodied as the 'habitus'. This embodied habitus manifests itself in the performance of what she termed 'practical knowledge' *without* prior planning or conscious thought – a habitus 'inculcated through the disciplines of the prison' (Caputo-Levine 2013: 169). In her own study with male released prisoners in the US, she considered the carceral habitus which develops amongst imprisoned men exposed to the probabilities and necessities surrounding the risk of interpersonal violence, in which 'bodily integrity must be maintained by constant vigilance' (Caputo-Levine 2013: 169).

The avoidance of violence has been widely observed as a structuring feature of prison life (e.g. McCorkle 1992, Hassine 2004) with a range of precautionary behaviours (such as avoiding certain places within the prison, carrying weapons) adopted and performed as 'routine', and an avoidance of identified activities which may serve to endanger (such as sleeping heavily). Whilst actual interpersonal violence *is* a genuine risk, just as Crewe et al. (2014) have argued for the UK, Caputo-Levine (2013) noted that it is the more mundane and banal aspects of violence which facilitate the carceral habitus – the verbal threats and abuse, and the insults which require appropriate response, without escalation to actual violence, to avoid the development of a bullying relationship. It is, she argued, the *threat* of impending violence, rather than *actual* physical acts of violence against the person, that create tension and fear and generate the precautionary behaviours of habitus.

She observed this embodied habitus in inmates' tendencies to display a well-developed upper body, and a particular manner both of holding the body itself, and of holding the body in space, which conveyed an ability to defend one's self, if necessary. There was also the adoption of a disconcertingly blank 'yard face' which functioned both as an expression of aggression, and to enable the individual to withdraw and assess a situation whilst concealing their intentions; and a particular hyper-sensitivity to physical space, leading individuals to operate vigilance by sitting with their backs to the wall when required to take a seat in a shared or communal area. An acute sensitivity to perceived disrespect, leading to snap reactions and sometimes physically and verbally threatening escalations, was also observed to characterise some inmates, as was an inability or reluctance to engage in small talk – both of which recall the risks inherent in verbal exchanges which carry the portent of threatened violence, in which relative positions in the social hierarchy must be maintained and the disclosure of information can lead to vulnerability. Caputo-Levine (2013) noted that there were differences in the levels to which inmates adopt these embodied strategies, which in her study were related to the nature of the facilities in which they were incarcerated (federal or state prisons, and specific prisons with particular reputations) and the length of time they had served, and also that some recognised their own semi-conscious adoption of the carceral habitus and tried to modify it.

Whereas for Caputo-Levine (2013) the 'carceral habitus' is the performance of practical knowledge *without* prior planning or conscious thought, for others, embodied strategies of incarceration are very much conscious and planned. For example, in work conducted with male prisoners in the UK, Sloan (2012: 403) highlighted the importance of personal cleanliness as a means of shaping identities, in that the cleanliness of the body itself, and the embodied routines and actions which enable cleanliness, 'symbolise how that individual wishes to be seen by others, and thus the important aspects of his identity than he values others seeing'. Although the prisoner may lack significant control over many aspects of his life, by maintaining personal hygiene, she argued, he actively and intentionally showed that he valued his body and wished to take care of it, demonstrating that he had control over this aspect of the self and how he was seen by others. Some prisoners noted that showering and the use of toiletries both structured prison time, and cleaned the body in ways that could 'relieve a degree of the infectious nature of the prison sphere' and 'enabled a degree of individuality to be placed upon the prisoner's body in terms of his own attributable smell or scent' (Sloan 2012: 404).

In their work on older women prisoners, Wahidin and Tate (2005) argued that imprisonment was experienced acutely corporeally by elder women prisoners, who complained about their physical deterioration during incarceration. Many of their respondents said that they had gained weight during their sentences, due to inactivity, poor diet, and overeating. Whilst these women bemoaned their weight gain as a negative marker of their imprisonment, resonant with Smoyer and Blankenship (2013), Smith (2002: 210) alternatively understood embodied practices of unhealthy eating as a means of wresting control over some aspect of prison life when much seems beyond personal agency. Thus, behaviours considered 'risky' or 'unhealthy' could also be understood as 'constituting a rational means of release, a way of coping, and of holding on to a sense of self'. The 'rationality of irrational action' (Graham 1984: 1989) enabled comfort eating to be interpreted as health-enhancing in this context, as a form of 'psychic survival', in resistance to the 'controlling' efforts of health promoters.

For other specific groups of prisoners, conscious strategies of embodiment shape and characterise experiences of imprisonment in unconventional ways. In their work with transgender women in men's prisons in the US, Jenness and Fenstermaker (2013) drew attention to the embodied activities in which transgender women engage in pursuit of gender authenticity, or what they call the 'real deal', in order to successfully negotiate the heteronormative and threatening environment of a men's prison. In a context in which (unlike on the 'outside'), transgender women cannot 'pass' as 'natural' or 'biologic' women – since all inmates in the prison are known to be male – transgender women committed to everyday embodied practices which communicated a desire to be read and treated as feminine. These strategies involved 'acting like a lady', through 'an intense preoccupation with bodily adornment and appearance as well as a deferent demeanour and a studied comportment' (ibid. 2013: 3), keeping good manners, accepting courteous treatment from male prisoners, and in some cases, acting 'like a lady' by urinating sitting down, and being seen to do so, in a cell shared with men. In extremis, however, a transgender woman could also engage in violence, 'standing her ground' (in the context of a suspended ladylike 'ideal') if necessary, to demand and secure respect 'as a woman'. Whereas there was friendly competition between transgender inmates for the attentions of men, transgender women reported benefitting from the care-giving provided by men, who recognised them as women within this context, and with whom they engaged in relationships. Men were accordingly seen as protectors and providers for transgender women, which Jenness and Fenstermaker (2013: 21) described as depending 'on the acceptance of the accomplishment of gender as indicative of a "natural" state, whether a male instinct to protect a woman, or the essential female qualities exhibited by transgender prisoners'. In this male prison context, they argued, it is the '*commitment of bodies* to act like, and be received as, "ladies" and "men"' (ibid. 22, emphasis in original) that dictated the gender dynamics between prisoners, and the nature of their everyday lives. In the harsh conditions of the male prison, they argued, the accomplishment of gender through the pursuit of the 'real deal' demonstrates the agentic power of embodiment asserted by Connell (2012).

Summary and Directions

Embodied experiences of, and emotional responses to, spaces of imprisonment, are amongst some of the most recent and most fascinating explorations of carceral space, both within carceral geography, and criminology and prison sociology. They bring to light contested and fluid notions of public and private in confinement, highlight the ways in which prisons are spatially differentiated, and draw attention to the micro-geographies of imprisonment, including those experienced at the scale of the confined body.

Consideration of the experience of incarceration in terms of its emotional register and the spaces in which emotional expression takes form, resonates directly with recent developments within human geographies of emotion and affect. A current debate within geography (McCormack 2008, Anderson and Smith 2001, Thrift 2004, Anderson 2006) revolves around a perceived distinction between emotion as 'embodied (subjective) experiences or significations', and affect as 'an impersonal set of flows moving through the bodies of human and other beings' (Simonsen 2012: 17). In her discussion of critical geography and phenomenology, Simonsen (2012) argued that an understanding of corporeality based on a non-dualist ontology of the body and its environment informed by Merleau-Ponty cut through this distinction, collapsing the binary of 'inner' and 'outer' into a double conception of emotional spatiality. In this conception, one 'side' is an *expressive space* of performance of emotions, where they are both practised and shown, and the other is *affective space* which is the space in which the body is emotionally in touch, and where emotions are not just actions to be expressed or articulated, but 'passions' by which we are possessed (ibid., emphasis in original). However, these two sides are not opposites – they form an active-passive circularity in which emotions are 'neither "actions" nor "passions" … they are both at once' (ibid. 18), and they are essentially relational.

Drawing on recent geographical engagements with emotions, embodiment and affect, carceral geography has the potential to build on this new theme of scholarship to bring a perspective of embodied emotions and affect as they are intricately connected to spaces of imprisonment. Although within emotional geographies there is already recognition that emotional management is influenced by socio-cultural circumstances (Davidson and Bondi 2004), the intensive management of emotional expression necessary within the carceral environment arguably presents a uniquely challenging context within which to consider emotional and affective geographies, but one which could also offer a mean to address question about the ways in which emotions are embodied in the context of 'less typical' everyday lives (Davidson et al. 2005: 5).

The work of Crewe et al. (2014) detailing the differential emotional expression possible in different areas of a UK prison, also suggests that this area of scholarship is one in which there is considerable potential for convergence between prison sociology, prison ethnography and carceral geography. Transdisciplinary dialogue on this topic could take further some of the observations made by Crewe et al. (2014), which would presumably be mirrored to some extent, within relevant contexts, in other carceral settings, and could begin to tease out the affective and emotional nature of a diverse range of prison spaces. Just as Crewe et al. (2014) have critiqued the relevance of the notions of 'frontstage' and 'backstage' which have long pervaded understandings of emotional expression during imprisonment, drawing on Simonsen's (2012) notion of an *expressive space* of performance of emotions, and an *affective space* in which the body is emotionally in touch, could be an entry point into a new means of framing the emotional and embodied experience of incarceration in ways which advance both criminological understandings of emotional expression, and geographical understandings of emotion and affect.

Chapter 5
Carceral TimeSpace

The majority of literature within the increasingly vibrant and growing body of work in carceral geography has thus far tended towards an overt spatial focus. Time has remained relatively overlooked, despite its central importance in the experience of incarceration, and its prominence in popular discourses about prison, that is, 'doing time'. Although carceral geography of course has an implicit awareness of the significance of time, inasmuch as it is time served inside of spaces of confinement that literally yields the research focus for much of this work, there is little critique of what time might *mean* in the carceral context, despite the increasing awareness and discussion of the innate variability of time within recent human geography.

This chapter therefore argues firstly that geographers working on and in spaces of incarceration could usefully consider and/or deploy some of the theorisations of time emergent within human geography, but also that in so doing they should not overlook the sophisticated understandings of time in prison already developed by criminologists and prison sociologists.[1] There is potentially a useful and productive discussion which could take place between human geographers working on spaces of incarceration, and criminologists and prison sociologists working on time in prison, and this chapter therefore seeks to initiate this process by bringing these disciplines into dialogue with each other, and by discussing specific studies within criminology and prison sociology which deal with time in prison.

In considering time in carceral space, carceral geographers might usefully look both to human geography and to criminology and prison sociology. While human geographers have grappled with ways to appropriately reflect and understand time within engagement with the spatial, criminologists and prison sociologists have dealt more explicitly with the issue of time and its experience, control and deployment, and increasingly demonstrate an awareness of the significance of space. The case for integrating the two is clearly articulated by criminologist Diana Medlicott (1999, 2001), who argues from her work on suicidal prisoners that 'Separating the time relationship from the place relationship is only justifiable analytically: in terms of the prison experience, the temporal and spatial aspects of existence … are experienced synthetically … Inmates' experience attests to the fact that the prison is a sophisticated time-place, where the temporal and the spatial characteristics are structurally productive of prison life and culture' (Medlicott 1999: 216).

---

1 Drawing on Moran, D. 2012. 'Doing Time' in Carceral Space: TimeSpace and Carceral Geography. *Geografiska Annaler B* 94(4): 305–16.

In response, this chapter reflects on the development of time geography based on the recent review of this work by Robert Dodgshon (2008), beginning with Hägerstrand, tracing the emergence of the various different ways in which a relationship between time and space has been negotiated within human geography, and drawing particular attention to the emergence of the concept of TimeSpace. It moves next to outline some key features of the discussion of time within criminology and prison sociology, paying particular attention to recent research into the experience of time in prison, and considering in some detail two specific studies. By revisiting the rich empirical material presented in publication of these studies, it discusses the way in which these prisoner testimonies could, in various ways, be interpreted as describing the TimeSpace of prison. It concludes by highlighting the potential areas of synergy between human geography and criminology/prison sociology in terms of the consideration of time.

Carceral Geography, Space and Time

There is considerable richness and variety within the geographical scholarship of incarceration, but a feature common to all of these studies is a primary focus on the spatial, and the associated marginalisation of the temporal. This is of course entirely understandable – this is a new field of study and geographers are still finding their way and exploring potential avenues of research. There is already an implicit interest in time, if only in that time spent in places of incarceration creates the experiences of these spaces which have formed the focus of much of the existing research. The importance of time within penal space also seems to be recognised; Sibley and van Hoven (2009: 205) appeal for fuller exploration of 'the relationships between prison architecture, the space–time regime, and correctional officers, on one hand, and the world of inmates, on the other'. However, in so doing they seem to reinforce a sense of separation between prisoners' experience and time, as if although prisoners may disrupt spatial control in the prison by exercising agency in creating and making spaces, their engagement with time is rather as a recipient, that time simply passes as it is 'served' but that prisoners do not engage actively with it.

As May and Thrift (2001: 1–2) have argued, drawing such a dualism risks viewing time as the domain of dynamism and progress, in contrast with space as the domain of stasis, 'excavated of any meaningful politics'. Massey (1994) points out that a dualism of space and time imposes unhelpful limitations on the potential theorisation of the spatial. In view of this, carceral geography needs to be cautious not to reproduce a one-dimensional focus on space which reflects an unhelpful dualism between space and time. Further, it should also avoid treating time and space in isolation, particularly in carceral space where, one might argue, they are unusually tightly bound together. In order to sketch out how this might be achieved, this chapter next considers two bodies of work towards which carceral geography might look for direction; human geography (time geography), and criminology and prison sociology.

Time in Human Geography

Recent work in human geography has stressed the innate variability of time, with a shift from objectified interpretations of time to a concern with relational forms of lived and experiential time, and a consideration of space and time as 'sticky concepts that are difficult to separate from each other' (Dodgshon 2008: 1). Hägerstrand's (1975, 1982) space–time analysis had at its core an assumption that people engage in the geography of everyday processes within a finite budget of time as well as within a defined framework of space: time as rooted in the lived circumstances of individuals, and experientially and spatially referenced. Focusing on time budgets, measured across days, weeks, or seasonal budgets of time, it was measured in standard clock or calendar time, and while it shed light on how individuals used time, it did not uncover the ways in which space and time were constructed or experienced. Subsequent discussions built on Hägerstrand's work in two significant ways; first by interrogating time's conceptual meaning, and second by paying attention to how its relationship with space is conceptualised, drawing attention in particular to time's relative rather than absolute nature. This trend runs in parallel with understandings of space, and facilitates the idea that neither time nor space are absolute containers or frameworks, but rather are 'bound up with how we see the world, particularly the spatial relationships between objects and the temporal relationships we perceive between events' (Dodgshon 2008: 1–2).

The understanding of the flow of time has also been explored in human geography, as well as being challenged by arguments that there are circumstances in which time does not flow – in which 'different times, past or future, simply occur in different places' (Dodgshon 2008: 9). In social science, 'time might be considered not to flow when no fundamental change occurs, as in the relative stasis of Braudel's (1980) *longue durée*, his enduring, inertial states when the changes that normally evoke time are absent' (Dodgshon 2008: 9). Considering the ways in which time does appear to flow, Dodgshon continues, the perception of forward flow stems from our 'most fundamental sense of time, one rooted in how we experience it biologically, experientially and culturally'. Imprinted on us biologically through our sense of ageing, and through a myriad of corporeal properties, we also sense forward flow through the interplay of the everyday with major life-course events, and derive a sense of passage of time through our own sense of becoming and a sense of difference which can arise from the everyday. The nature of time flow can further be distinguished as continuous and regular (as in the even measure of clock time and work discipline), or continuous but uneven (in the case of experiential time and its innate variability of flow). Flow can also be disrupted by disconnected or disjunctive moments.

Human geography has positioned itself at the intersection of debates over the relationship between time and space; between the growing primacy of the spatial in social theory, and the increasing call for the rediscovery of time in the study of space. Human geographers have integrated these debates, with writing such as that by Thrift (2000), May and Thrift (2001) and Massey (1999, 2005) giving rise to the discussion of space–time or TimeSpace rather than space and time (Dodgshon 2008: 11).

Dodgshon (2008: 11–12) observes that human geographers have handled the relationship between time and space in four distinct but related ways. The first two of these distinguish between space and time and emphasise one at the expense of the other, as timeless space or spaceless time, in studies that suppress time, or which deny an explicit spatial framework to a narrative process. In the third, some approaches regard space and time as interdependent, but distinct one from the other; by spatializing time (for example contrasting different places as experiencing different stages of linear development) or temporalizing space (such as developing a history of a single place) to make each accessible to the other. Finally, some approaches seek to erase the differences between space and time until they cannot be analytically separated from each other. These approaches consider space–time or TimeSpace (Wallerstein 1998, May and Thrift 2001), 'dealing with socially embedded space and time that emphasises their interdependency, such that space is inextricably bound up with society's experience of time and time consciousness' (Dodgshon 2008: 12).

As Middleton (2009) points out, time-geography has been critiqued by feminist and cultural geographers, who stress that time is a resource to which not everyone has equal access. These critiques mirror those which stimulated the proliferation of work in human geography on mobility as a means of access to space, and which draw on a relational approach to space and social inequality, seeing mobilities as spatial means of creating, maintaining and deepening inequalities (e.g. Larsen et al. 2007, Cresswell 2008, Manderscheid 2009, Jensen 2011). Incarceration is clearly understood by geographers as the denial of mobility and access to space; Philo (2001: 474) described 'an institution such as a prison necessitates a definite spatial concentration of population, a collecting together of a largish mass of human beings in a small and in this case clearly confined, delimited and set-apart area', and much of the current work in carceral geography engages with the implications of this situation for those held in these confined spatial concentrations, and increasingly with the nature and significance of mobility in this context (e.g. Moran et al. 2012, 2013). What it does not yet do is consider access to time, the signification of time, or the nature of the relationship between time and space.

A focus on time in carceral space should clearly draw upon the wealth of work within time geography, summarised above, but it would also benefit from careful consideration of the work of criminologists and prison sociologists. Whilst geographers have debated appropriate ways to reflect and understand time within engagement with the spatial, for prison scholarship the focus has been more explicitly on the temporal: time and its experience, control and deployment.

This is not to argue that space has been overlooked by criminology and prison sociology, however. Despite a guarded acceptance of Foucault's (1979) assertion that the design of prison spaces to enable actual or perceived constant visibility and surveillance has an effect on the behaviour and control of prisoners (Alford 2000), the spaces of prisons has, until relatively recently, remained under-researched in this field (Canter 1987, Ditchfield 1990, Fairweather and McConville 2000, Marshall 2000; Hancock and Jewkes 2011).

Time in Criminology and Prison Sociology

Time as an element of study within criminology and prison sociology has taken a number of forms. Researchers have considered time as a 'given' constant, as an axis of differentiation, for example through longitudinal studies looking at changes in various phenomena such as; imprisonment rates, levels of overcrowding and prisoner welfare (e.g. Jacobs and Helms 1996, Stucky et al. 2005), individual prisoners' experiences (e.g. Zamble 1992), and prisoners' adjustment to incarceration (e.g. Warren et al. 2004, Thompson and Loper 2005), considering all of these over time. They have also explored imprisonment as a discrete period of time which is thereby distinguished from the rest of prisoners' lifecourses (e.g. Pettit and Western 2004), and have studied the variation in, and effect of, different lengths of prison sentences (Aebi and Kuhn 2000), and the particular issues associated with the passage of time in terms of the incarceration of people at different stages in the lifecourse, for example younger people (e.g. Biggam and Power 1997, Cesaroni and Peterson-Badali 2005) and older people (e.g. Aday 1994, Codd 1998, Howse 2003, Crawley 2005, Rikard and Rosenberg 2007).

Within the prison itself, time and imprisonment are of course integrally linked, as 'time is the basic structuring dimension of prison life' (Sparks et al. 1996: 350, cited in Cope 2003: 158). For Foucault (1979), for example, control in prison was exercised through time-discipline, which limited inmates' abilities to make decisions about their activities. If carceral geographers have tended to identify most readily with the spatial in the work of Foucault, in criminology and prison sociology the temporal has also proved significant. The perception of time in prison has become a prominent research focus, in that although prison time tends to be treated as being externally controlled, variously by a judiciary who sentence inmates for a certain length of time, by prison officials and others who decide on parole entitlement, and on a daily basis by guards who enforce flow control and the penal regime, prison scholarship recognises that the experience of time is both personal and variable. It has explored inmates' abilities to cope with time, related to their age and their awareness of self-deterioration during sentences (Farber 1944, Cohen and Taylor 1972), within a body of literature which considers time in relation to prisoner vulnerability (Medlicott 1999). Although space is less prominent in this analysis, recent work has explored how the embodied experience of carceral space enables prisoners to 'renegotiate the disciplinary techniques of time' (Wahidin 2002: 180).

The remainder of this chapter draws upon recent research in this broad area, and specifically upon the empirical materials generated and published by Nina Cope, and by Azrini Wahidin and Shirley Tate, within the field of criminology/ prison sociology (Wahidin 2002, 2004a, 2004b, 2006, Cope 2003, Wahidin and Tate 2005). Cope's work focused on the experiences of male young offenders aged between 16 and 21, serving sentences in a young offender's institution in the UK. Wahidin's (and Wahidin and Tate's) work created data based on the experiences of women aged between 50 and 73, also serving sentences in the UK. In both cases data were generated via semi-structured interviews with inmates within prison institutions, and in each case the specific methodologies employed, and the ethical considerations attendant to the research, are explored in detail in the publications listed.

Although time as broadly conceived of within the carceral context is 'clock' time, Wahidin (2002) and Cope (2003) acknowledge that prison provides a context for multiple temporalities. Wahidin (2002: 182–3) in particular explores how conceptions of time in prison are structured and controlled, and interrogates the meanings of time in prison, the ways in which these become dispersed and contested, and the ways in which this affects prisoners' identities, bounded both spatially by physical walls, and also by 'time-space walls on all sides' (Binswanger 1984, cited in Wahidin 2002).

Through their work in this area, these scholars have developed a sophisticated understanding of prison time which mirrors many of the conceptualisations underpinning work in time geography; in particular highlighting the overlapping temporalities which exist within carceral space, such as the externally imposed clock time which measures sentences in days, weeks, months and years, and the experiential time as experienced by individual prisoners, who variously sense stasis (with time seeming to stand still while they are incarcerated through the daily repetition of penal routines), who perceive time to flow more quickly outside the prison than inside of it (as events in the lives of others seemingly pass them by), or who observe the passage of time biologically (through their own embodied processes of ageing and attendant physical deterioration). They have additionally considered time as embodied, or as Wahidin (2002: 182) puts it, 'body time', and have discussed ways in which prisoners seek to wrest some form of control over time, through the deployment of various resources and tactics (Wahidin 2002, Cope 2003, Wahidin and Tate 2005).

It is here, in the consideration of the embodied experience of time in carceral space, that there seem to be synergies with human geography. Geographers engaging with carceral space could fruitfully draw upon scholarship in criminology and prison sociology both to understand the importance of control over time in carceral geography, and to reflect the relevance of time-geography to rematerialised human geography (Schwanen 2007).

TimeSpace in Prison

'We access all time through the portal of the present or each successive now' (Dodgshon 2008: 7). Key to our experience of time is the widely accepted understanding that although we apprehend that past, present and future are different, they do not strictly have an existence outside of the present, and cannot be defined in any way other than in their relation to it. Whatever we think about the past and future, we do that thinking or have those impressions in the present, and in the context of what the present is like. The present therefore becomes transformed from a mere instant or passing moment, into an extended present, in which there is a sense of continually becoming, with each new moment adding to the past through the work of memory. Although precise interpretations of the nature of the present differ (Bergson 1911, Merleau-Ponty 1962, cited in Dodgshon 2008: 7, Husserl 1983), the core belief that all time is accessed through the present or 'each successive now' is shared. As criminologists and prison sociologists have recognised, the very fact of being in prison, mediated by inmates' age, gender, the length of their sentences and any previous experience of incarceration, changes the way in which prisoners experience time. The 'now' of incarceration, taking the form of an extended present, in which each new moment of imprisonment is added to the past as a moment of memory, shapes prisoners' thoughts and feelings about their past and their future, and their sense of the passage of time. An elder woman interviewed by Wahidin (2004b: 81) describes this situation:

> ... each day your life slips away from you a little bit and then one day you wake up in the morning and it's all gone and that's the worst day of all, when everything before has gone. I was writing, it was like a diary effect, I was writing about present day and past and it was all mingled in together, it sort of made you cry and laugh at the same time, and that's what happened to me along that way. It all just slipped away. You can't keep it, you can't hold it.

The nature of that now is not just temporal, however. It comprises both the temporal and the spatial; the time and the space – the TimeSpace – of incarceration, and is bound up with the corporeality of the individual whose now is being experienced. The embodied experience of time is inextricably bound up with the embodied experience of space, and vice versa. In other words, it is not just the fact of an individual's being in the present of a prison sentence, but of their being physically present in a carceral space, that determine the nature of the now which colours their perception of past, present and future and the passage of time.

In their discussion of experience in prison for young men and elder women, Cope (2003), Wahidin (2002) and Wahidin and Tate (2005) present empirical materials which describe carceral TimeSpaces. Their respondents discuss the varying temporalities within carceral space; between clock time and experiential time in the perception of stasis, the nature of the embodied experience of carceral

TimeSpace and the way in which the embodied experience of time through ageing varies between different stages of the lifecourse. The following sections will explore these in turn.

Clock Time and Stasis

As Middleton (2009) points out, time geography has tended to treat time as a resource to which everyone has equal access, and has tended to overlook the power relations embedded within the experience of time and the way in which agency is deployed to exert actual or perceived control over experiential time. The empirical material generated by Cope and Wahidin reflects just such deployment of agency, exploring the different speeds at which prisoners' experiential times flow, and the strategies employed to make time seem to flow faster, in response to a sense of temporal stasis, in which time does not flow because no fundamental change occurs. In Cope's study (2003) young male offenders describe the repetitive, rhythmical sameness of prison routines (and the absence of different experiences) leading inmates to view their sentences as time not moving on. One said (cited in Cope 2003: 161):

> I've got used to it [the regime in the prison], I mean in other jails they do association during the day and bang up at night [locked in cells], this jail it's association in the evening. When I was in my cell all day I was bored and used to sleep, but now I'm working … it's [prison] time wasting 'cos you don't do nothing, you do the same thing every day. You get up, go to work, eat, have a shower, work, association, bed, then get up in the morning and it's the same again. The same thing every day.

Presenting the words of elder female inmates, Wahidin and Tate (2005: 65) found similar opinions:

> Time has stood still while I have been in prison. Time has stood still in that everything goes on and on in the same repetitive way. It is as if the twenty years could have all been fitted into one year.

Time in prison is on the one hand straightforwardly clock time, in that prisoners are given a sentence of time to serve, measured in years, months, weeks, and days; however on the other hand clock time becomes only one of a number of overlapping temporalities within carceral space. Prisoners are acutely aware of clock time, as one of the elder women in Wahidin's (2002: 183) study described, breaking down the days into hours, minutes and seconds as she watched the clock ticking.

[prison time is] quite different because you're watching the clock inside prison whereas you don't necessarily look at the clock on the outside. I mean the clock was there and you might glance at it, but here clock time means everything.

But whilst there is a constant awareness of clock time, its flow does not feel continuous and regular to prisoners. According to Cope (2003) they try to assert control over it, performing certain activities which make the flow of time seem faster, or to avoid it feeling slow. For example, the young male inmates in her study used cannabis in order to sleep, offering them a way to allow time to pass without having to experience every waking moment of it (cited in Cope 2003: 167):

With cannabis you can smoke it at night-time and it makes your head go down, it makes me relaxed and makes me fall asleep. The way I look at it is it makes time go faster, cos there are times when you can't sleep and that ...

Similarly, some of these young men opted not to receive visits so that they would not have specific events to look forward to, since they found that this expectation and anticipation made time pass more slowly (cited in Cope 2003: 171):

I used to be getting regular visits, but recently I haven't been sending out the VOs [visiting orders], 'cos I know if I'm waiting for a visit, time goes slow, so I leave it for a little while and send them out after a few months and have some for a few weeks or something.

An elder woman (cited in Wahidin and Tate 2005: 67) described a similar tactic in marking off days in a calendar:

Oh yes, I was crossing them off and then I was leaving them to the end of the month. I've got my calendar and I thought to myself now I've left it, I've marked half of March off. Tonight I'll mark the other page off and turn the page over to April and see if it goes any quicker that way, by doing it at the end of each month instead of doing it day by day you know.

These two groups of prisoners shared a sense of the variation between clock time and experiential time, and of the stasis of time in prison. Where they diverged was in the embodied experience of time; the perception of passage of time varied according to their respective stage in the lifecourse, but even though they experienced this differently, they still deployed tactics of control.

The Embodied Experience of Time

When discussing the passage of time in terms not of clock time and how it can be made to flow faster or slower, but of the flow of time imprinted biologically through a sense of ageing and its myriad corporeal properties, the views of the elder women and young men differed markedly. They saw things differently, in the embodied context of their stage in the lifecourse. There were particularly stark differences in the ways in which these two groups of inmates viewed the future, with age proving a significant factor. Wahidin and Tate present compelling evidence that the passage of time in prison was felt corporeally by elder women prisoners (cited in Wahidin and Tate 2005: 70):

> Inside I don't feel old. But my body feels old. Health-wise – I have no energy and my bones ache all the time. When I was at home on a morning I'd be as bright as a button. Looking forward to the day. But here, it's such an effort. You seem to be dragging your body around all the time. You are conscious of your body. You know, it feels heavy all the time. Your heart is heavy all the time. Your feelings are all heavy – there is no light-heartedness.

> [My health has] deteriorated badly. There is nothing else I can say. The sheer boredom in prison. I was a healthy person at 9 stone, which is 1 stone overweight than I should have been. I was a healthy person; I was active. I played a lot of games, badminton and tennis. When I came to H Wing I went up from 9 stone to 16 stone. That was sheer comfort eating. Now, I have rheumatoid arthritis, which has meant that when I have gone out and all the things that I have dreamed of doing while all these years inside, I just can't do now.

Whereas the elder women seemed to feel the weight of time as a physical burden, and described their relationship to past and future in terms of what they had been able to do in the past but could not do now or would not be able to do in the future because of physical deterioration, the young men's youth seemed to 'have made suspending time and discarding the years off their lives easier to rationalise', in that they regarded the sentence to be a 'temporary marking of time' (Cope 2003: 166) and focused positively on their youth and life ahead on release. Cope (2003: 164) found that although 'time passed from day to day, this daily passing of time was not seen by the inmates to impact on their physical development, ageing and maturation'. Specifically, she found that these young men described the 'sameness' of prison to imply that they did not 'need' to mature (cited in Cope 2003: 165):

> I won't grow up, time stops dead in jail, don't it? You're doin' the same thing every day, livin' the same life every day, so there is no need to grow up. When I get out I'll still be a seventeen year old. I'll be thirty odd when I get out, but still doin' things that a seventeen year old would be doin'. I don't think I've grown up, I think I've got more clued up, you don't grow up mate, it's all fun and games in't … time stops dead [in jail].

As Cope describes, they took control over time by separating time in prison from time on the outside, as if to protect their external identities from time served in prison (cited in Cope 2003: 165):

> Say I'm gonna live to be seventy and I do four years of this prison sentence yeah, when I leave here I'll be twenty three, no, I'm gonna be nineteen, more mature than most nineteen year olds. They're just prolonging my life 'cos I will be seventy four rather than seventy.

Although the elder women seem to be comparatively disempowered through the ageing process, there are still instances in which Wahidin (2002: 190) stresses that some deploy tactics which temporarily enable them to 'control the prison gaze, the violation and the situation'. In the context of a routine and period strip search, one woman described her own strategy (cited in Wahidin 2002: 190, emphases in original):

> The [first time you] strip off. It is a really degrading thing. For a long time it used to bother me. If I knew I was due for a strip search, I'd think "oh God, they'll be coming in a minute. They'll be coming and I'll have to take my clothes off". You see how I have changed. I've changed and I've become stronger and I think to myself, "I whip it off". And they say "Don't take your bottoms off before you put your top back on". I think, "well blow you because what's the difference I'm being stripped stark naked". So I just throw everything off now. But it's my bravado. If you do it like that really quickly I feel as if they take a step back. They are more embarrassed than I am.

Another woman described a similar tactic through which, as Wahidin and Tate (2005: 72) argue, she used the body to symbolize the undermining of discipline:

> I stood for a moment in this white lacy body and thought I'll just let them see how well I look for the age of 68, and then I peeled it right off and handed it, twirled around, carried on laughing and that is the first time I've ever dealt with a strip like that. Whereas before I used to get really distressed and everything.

Wahidin (2002: 192) argues convincingly here that the material body almost becomes a medium through which power operates and functions, and that the body is not passive but rather is interwoven with and constitutive of systems of meaning, signification and representation through which elder women negotiate the penal gaze.

Conclusion: Carceral TimeSpace

These fascinating and compelling empirics generated by Cope and Wahidin (Wahidin 2002, 2004a, 2004b, Cope 2003, Wahidin and Tate 2005) demonstrate the sophistication of criminologists' and prison sociologists' engagement with the experience of time in prison. Bringing them into dialogue with carceral geography reveals the potential for some interesting synergies.

The overt spatial focus of recent work in carceral geography renders it almost a study of 'timeless space' (Dodgshon 2008: 11), in which the temporal dimension of incarceration is neglected in favour of the spatial. This spatial focus needs to be balanced by a consideration of time in carceral space, and there are a number of ways in which this could be done. Taking a lead from Hägerstrand's time geography, and regarding space and time as interdependent but distinct, incarceration could be viewed simply as a geographical 'space–time fix' (Schwanen et al. 2008), a constraint that binds an activity (in this case imprisonment) to a specific place (a place of incarceration) and a specific moment in chronological time (during a prison sentence), and the degree of fixity, and the level of prisoners' fixity in both time and space and the ways in which these vary by context and individual could be investigated. Alternatively, and arguably more promisingly, carceral geography could recognise that time and space are co-constitutive, and in so doing build upon the rich vein of work within criminology and prison sociology which theorises time as experiential, and which opens a space for the discussion of TimeSpace in the carceral context. In privileging neither space nor time in the carceral context, but instead considering carceral TimeSpace, a number of potential avenues for future research present themselves.

First, such a perspective would bring together parallel concerns in carceral geography and criminology/prison sociology over agency in the carceral setting. In carceral geography there has been considerable attention paid to the tactics deployed by prisoners within carceral space, with studies highlighting the reclaiming of space (Dirsuweit 1999), the production and reproduction of spaces (Sibley and van Hoven 2009, van Hoven and Sibley 2008), and the personalisation of space by prisoners (Baer 2005), with attendant theorisation of these practices as in some way representing an exertion of a form of control over space, in tension with the theorisation of prison space as regulated by Foucauldian disciplinary power. The examples from prison scholarship presented here suggest that the same types of processes take place in relation to prisoners' relationship with time, with tactics deployed to create a sense of control over the passage of time, and these parallel sets of observations of space and time might usefully be brought together through TimeSpace.

Second, a focus on carceral TimeSpace could provide an informative example for lifecourse research, where a lifecourse approach involves 'recognition that, rather than following fixed and predictable life stages, we live dynamic and varied lifecourses which have, themselves, situated meanings' (Hopkins and Pain 2007: 290). Geographers have tended to view incarceration as the experience of a carceral space, both in terms of the individual's movement into and out of that space and their experience within it, as well as the physical manifestation of the penal institution in space, whereas criminologists have arguably tended to conceptualise incarceration as a period of prison time (e.g. Pettit and Western 2004). According to the data generated by Cope (2003) and Wahidin (2002, 2004a, 2004b, also Wahidin and Tate 2005), the stage in the lifecourse matters in understanding the experience of incarceration. Considering incarceration as a TimeSpace within the lifecourse would draw together these two approaches, providing a theoretical framing for conceptualising incarceration co-constitutively as spatial and temporal.

Third, and relatedly, consideration of the embodied experience of carceral TimeSpace is critical to understanding the experience of incarceration. As Cope and Wahidin have shown, when prisoners talk about carceral TimeSpace they do so by reflecting on their past and future, and by describing their experiential time as experienced bodily. To understand the now of carceral TimeSpace, through whose lens both past and future and the passage of time are viewed, is also to appreciate the embodied nature of this now. Although criminologists and prison sociologists like Wahidin have incorporated consideration of the incarcerated body into their work, there is still scope for carceral geographers to participate in the interdisciplinary inquiries 'involving 'embodiment' as it relates to the spatiality of bodies and the affective and performative aspects of living in and making spaces and places' (Butcher 2012: 91), but rather than to limit themselves to the aspects of living in 'spaces and places', to consider the embodied spatiality and temporality of carceral TimeSpace.

Finally, and most fundamentally, TimeSpace would seem to be a useful conceptual tool for advancing understanding of what incarceration is, which cannot adequately be achieved by focusing on either on time, or on space, in isolation, or even, perhaps, in combination. This assertion resonates with Merriman's (2012: 24) recent questioning of the a priori focus on space-time or spacing-timing within human geography. He argues that this focus 'overlooks the importance of other registers, apprehensions, engagements and movements that appear to be important for understanding the unfolding of many events', and suggests that rather than privileging space and time in the conceptualisation of location, position and context, we might instead consider movement, rhythm, energy, force or affect as registers of equal importance. As the empirical material referred to in this chapter has demonstrated, the consideration of carceral time and space, or TimeSpace, is not an end in itself; it arguably operates here as a means of uncovering and understanding those other registers, such as rhythm, force or affect, in the experience of incarceration.

PART II
Geographies of Carceral Systems

Chapter 6
Geographies of Carceral Systems

This chapter explores the spatial or distributional geographies of carceral systems. Just as levels of incarceration vary widely in a global context (see Table 2.1, Chapter 2), under one criminal justice system, the distribution of carceral facilities across space is neither even, nor entirely attuned to population, or crime, distribution. However, where places of incarceration are located has considerable significance for the communities which surround these places, for the communities from which prisoners come, and for the incarcerated themselves. Discourses around the siting of prisons have tended to focus on the relationship between prisons and the communities which surround them, irrespective of whether or not these are the communities from which the prisons' own inmates have come. A distinction is drawn between them here, since, as recent research has demonstrated, the specific location of prison facilities frequently means that they are populated by individuals whose domicile is some distance away, and whose community and family ties are stretched to breaking point by their incarceration. Key to these geographies is the process of prison siting, which is a highly contested, politically (and increasingly economically) charged procedure, with far-reaching consequences for communities and prisoners alike.

Although the rationale for the location of new prisons may change, once built, prisons are spatially fixed until decommissioning and closure. In long-standing carceral systems, at any one time, the geography of the system will represent a complex combination of previous, as well as current, logics of prison siting, which themselves will have operated within prevailing cultural norms and traditions, economic circumstances, political prerogatives and notions about the purpose of imprisonment (Lynch 2011). Historically, the rationale for the siting of prisons may change – for example, in the case of the UK there was a shift in the twentieth century from a preference for building imposing prisons in metropolitan centres to hiding them away in more extra-urban locations where land rents were cheaper and opposition more dispersed. Nevertheless, the prison estate, especially in countries with an expanding prison population, tends to show a high degree of inertia. Previous logics of prison siting which have become physically manifest, sedimented and institutionalised through prison building may no longer fit the intentions and imperatives of prevailing penal philosophies, leaving criminal justice systems with penal facilities in suboptimal locations, creating and sustaining problems for communities and prisoners. Prisons are expensive and controversial both to build and to close, so they tend to remain where they are.

Mindful of these issues of context and rationale, this chapter unfolds as follows: first with an overview of research into prison siting which highlights the so-called 'NIMBY' response of communities resisting the proximate location of new prisons. Next it moves to discuss the effects of the 'prison boom', the recent rapid expansion of penal systems, especially in the United States, and the associated diversification in attitudes towards prison location, as well as discussing the embeddedness of long-standing prison institutions within local communities to which they provide employment and in some cases, a sense of identity. Whilst much of this work originates in criminology, sociolegal studies and economics, geographers have recently made significant contributions to these debates. Finally it considers the impact of prison location for prisoners themselves and for the communities from which they come, drawing on recent research into the impact of distance from home on experiences of incarceration. Although the great majority of research in this broad area has focussed on the United States, largely in relation to the on-going debate over the US prison boom and the location of correctional facilities in rural areas, the chapter also draws upon studies from the UK and the Russian Federation.

Not in My Backyard: Resistance to Prison Location

Spatial geographies of incarceration have been inspired by concern for the impact of the distribution of places of incarceration on the communities which host or surround them, and they are frequently, albeit often implicitly, in dialogue with ideas of the 'total institution' Goffman (1961), drawing attention to the permeability of the prison wall in terms both of the perceived leaching out of negative influences from the prison into surrounding communities, and the anticipation of positive impacts on local economic development deriving from prison siting. Much of the literature examines the 'not-in-my-back-yard' (NIMBY) attitude of many host communities, (see for example Sechrest 1992, Martin 2000, Thies 2001, Combessie 2002), with research on prison siting and the relationships between prisons and local communities highlighting the traditionally assumed opposition of communities to location of prisons close by.

   Negative attitudes towards prison location are often recognised and addressed as part of public consultation processes ahead of prison construction. For example, in an independent report produced as part of one of these consultation processes, for the Matanuska-Susitna Borough, in the US state of Alaska, ahead of public hearings about a planned prison, Groot and Latessa (2007) observed that they fall into two main categories – concerns about safety, and about financial wellbeing. Safety concerns usually comprise anxieties over increased crime rates, prison escapes, and the relocation of (presumably 'undesirable' elements from) prisoners' families. Associated financial concerns include worries over falling property values, and costs of infrastructure development. Taking these in turn, the evidence for them is mixed. Fears of rising crime rates are amongst the most commonly cited reasons for opposing new prison plans (Myers and Martin 2004), although the reasons why building a prison should be expected necessarily to bring crime to an area are unclear, since numerous studies have shown that communities hosting prisons do not experience elevated crime rates (Abrams and Lyons 1987, Sechrest 1992), and as Thies (2000) has argued, they may in fact experience lower rates of criminal behaviour, perhaps because of the local presence of criminal justice personnel. Fears about the crime rate implications of visitors coming to the prison, presumably based on assumptions of criminality-by-association, also seem to be unfounded, as do fears that prisoners will be released locally (rather than being transported to their home area) and will commit offences immediately after release (Schichor 1992). Concerns over escapes seem to be associated with media coverage of such events (Thies 2000), but escapes are rare, in the US at least. Finally, some local residents harbour fears that prisoners' families (again presumed criminal by association) will relocate to be nearer to their incarcerated loved one, and this anxiety is compounded by the associated fear of overburdening local civilian infrastructure such as schools and hospitals (Groot and Latessa 2007). In fact, prisoners' families very rarely relocate (Sechrest 1992, Schichor 1992). Associated financial concerns over prison location include negative impact on property values, assumed to occur with the location of a prison as a 'Locally Unwanted Land Use' (LULU). However, evidence here is again mixed, with some studies showing that property values decline only when local opposition is particularly vocal and attracts media coverage, which in turn alerts potential homebuyers and investors (Abrams and Lyons 1987). Concerns are also expressed over the costs of infrastructure development, in order for local facilities to cope with the additional strain placed on them by a new prison (Blankenship and Yanarella 2004). This anxiety does seem well grounded – although rather than representing a burden *after* the prison has been built, infrastructure development tends to be undertaken as an investment in order to attract a prison to an area.

Although focus has traditionally been on structural change associated with prison siting, more recently questions have also been posed about the response of local communities to the *aesthetic* appearance of prisons, and the importance of prison architecture in the 'acceptance' of prison siting close to existing communities. As Armstrong (2012) found in her study of prison siting in Scotland, members of a local community described a proposed prison as a 'monstrosity of a building', a 'massive edifice', a 'monumental monstrosity lit up at night'; argued that the prison's design was seen to be 'out of keeping' with the perceived nature of the surrounding area, and that a 'large, unambiguously manmade, permanently lit monolith' would irrevocably change the character of the local area (ibid. 17). There is more than a suggestion here that the aesthetic appearance of prisons is of considerable, yet under-explored, importance for local residents.

Profit through Punishment?

There is increasingly an alternative to the NIMBY response, that of the notion of 'profit through punishment'. As Martin (2000) pointed out, the consideration of prison location as a form of LULU, alongside power stations, waste dumps, and other necessary but 'unwanted' facilities, significantly coloured the early literatures on prison siting, and he argued that the lack of systematically collected data on community responses to prison siting contributed to a sense of strong resistance to, and negative attitudes about, the introduction of prisons to local communities. His work on the siting of a prison in the US state of Pennsylvania identified that community responses were at worst neutral, with residents' views closely mirroring objective indicators of the relatively minor impact of prison location indicated in the literatures cited above (low risk of escapes, little impact on local crime rate etc.). Martin's work, and the timing of it, is significant in that it suggests that in his research context at least, community views of prison siting were much more positive than the previous 'accepted wisdom' of NIMBYism may lead us to believe. We might conclude from this work either that community views have either always been relatively mild, just rarely solicited in this way, or there has been an amelioration of opinion as the US carceral estate has expanded in the so-called 'prison boom' (Spelman 2009) and the perception of prisons as potential stimuli for economic growth has gained traction.

There has therefore arguably been a shift from fear and anxiety over the siting of prisons, to a situation of demand, especially on the part of small rural towns in the United States, for the building of prisons as stimuli for local economic development and employment (Eason 2010, Che 2005, Cherry and Kunce 2001). For example, Cherry and Kunce (2001) found that in California, policymakers located 'inferior' public facilities, such as prisons, in less prosperous neighbourhoods, partly because there was anticipated to be less 'NIMBY' protest there than in more prosperous areas, and, unable to attract private commerce, these areas may be more willing to 'accept' opportunities 'discarded' by others. However, just as the evidence for negative prison effects is tenuous, the effectiveness of prisons as growth stimuli in this regard is also highly debatable (Bonds 2006, 2009).

As Huling (2002) has noted, prison officials commonly attempt to convince rural communities of the economic benefits of prisons; local officials sponsor town meetings where prison officials and their supporters are invited to extol the virtues of prisons to communities, local newspapers are filled with articles reporting grand claims for economic salvation, and flyers are distributed to local coffee shops, general stores and mini-marts. The purported benefits are described by a California Department of Corrections official who stated that:

> Prisons not only stabilize a local economy but can in fact rejuvenate it. There are no seasonal fluctuations, it is a non-polluting industry, and in many circumstances it is virtually invisible ... You've got people that are working there and spending their money there, so now these communities are able to have a Little League and all the kinds of activities that people want. (Cited in Huling 2002: 198)

US states also compete with each other for federal prisons; in 1997, Pennsylvania offered up to 200 acres of prime state-owned farmland in rural Wayne County to the Federal Bureau of Prisons for $1 (ibid.). In this competitive process, bidding wars can escalate costs to local communities, as they vie to offer the most attractive lures, donating land, expanding infrastructure and providing subsidies (Tootle 2004, Hoyman 2002). However, there remains very little empirical evidence to support a connection between prison siting and community benefits. Although as Glasmeier and Farrigan (2007) noted, the opening of a 400-employee prison facility in Appleton, Wisconsin brought new income to the community, and the prison became the largest taxpayer in the county, other examples, such as in Pender County, North Carolina, demonstrated that the economic impacts of the prison development boom on persistently poor rural places, and rural places in general, were limited.

Whilst Glasmeier and Farrigan's (2007) analysis suggested that prisons *may* have had a positive impact on poverty rates in persistently poor rural counties, they remained unconvinced that the prison development boom resulted in structural economic change in persistently poor rural places. More likely, they argued, is that positive impacts were attributable to spatial structure; that is, to the mere existence of a new prison operation in a rural place rather than the facility's ability to foster economy-wide change in terms of serving as an economic development initiative. Hannan and Courtright (2011) noted that although the magnitude of the actual economic impact of prison siting may be important, it is often difficult to quantify; recent studies have found little evidence for either a positive or a negative economic effect on local communities (Dickinson 2003, Hooks et al. 2004, King et al. 2004). Although prisons might be assumed to bring investment and employment, King et al. (2004) found that local residents seldom found work in the correctional facility since they did not have the requisite skills; that the correctional officers employed did not live in the local area; that few multiplier effects developed, and that inmate labour started to displace local unskilled workers.

The recent economic downturn has revealed the full extent of the economic misadventure of prison building. Norton (2014) observed that communities in the Northern Adirondack Mountains in New York State are realising the implications of the 'failed economic development project of prisons'. Genter et al. (2013) found not only that prisons did not improve job prospects for inhabitants of host communities, but that the privatisation of prisons in the US has actually *impeded* local economic growth. Direct mechanisms such as the significant reduction in both prison staffing and salary levels in privatised prisons (and in public prisons which have tracked these changes), have depressed local development. However, they also drew attention to indirect mechanisms such as the opportunity costs in prison recruitment investments and subsidies, which may have been considerable for private prisons (Price and Schwester 2010). Such investments often diverted resources away from education, childcare and other public programmes which have a more clearly defined and documented impact on local employment and economic growth. In other words, not only have prisons themselves failed to deliver growth, but competing for them has unintentionally stymied the development

which may otherwise have taken place, leaving host communities doubly exposed
to the effects of prison privatisation and economic recession.

The marginal nature of economic benefits accruing to host communities from
prison siting arguably make the *perception* of economic impact on the part of
local residents critical in shaping prison-community relations (Turner and Thayer
2003). Through a study of residents of rural communities in Pennsylvania, US,
Hannan and Courtright (2011) found that perceptions of the impact of a prison
were less positive on the part of respondents living closest to the facility – perhaps,
they argued, because such close neighbours were well placed to judge the facility's
economic relationship with the community. Alternatively, they suggested that
these close neighbours, particularly those who were resident in the area when the
facility was constructed, were better able to 'compare reality with expectations
at the time of siting, and, in short, separate reality from expectation' (ibid. 62).
On the basis of these findings, they echoed Carlson's (1991) earlier work by both
urging the promotion of realistic expectations of prison location, and emphasising
the importance of correctional facilities' employment of local labour, and the local
procurement of goods and services (Hannan and Courtright 2011 and Courtright
et al. 2009).

Whether or not US rural towns vie for or agitate against prison siting, and
whatever the objective impact of prison siting on these rural communities,
the trend for new prisons to locate in extra-urban places brings with it a set of
controversial implications. The movement of prisoners in the US from inner cities
to rural areas, in what is termed the 'prison industrial complex' (PIC), is argued
by Huling (2002) to entail three systemic problems. Firstly, prisoners 'count'
in the US census where they are imprisoned rather than where they used to live
pre-incarceration; secondly the US has nearly doubled its prison population since
the 1990s; and thirdly the rural prison boom during the 1990s therefore means a
substantial transfer of resource from urban to rural America, since federal funds
are tied to population counts. By 2012, more than 60 per cent of the US's 1,600
prison facilities were located in nonmetropolitan communities (Eason 2012: 274).
Inner-city communities, 'home' to large numbers of prisoners, lose state funding,
and the prisoner 'share' of the nearly $2 trillion in federal funds distributed
nationwide is redirected to the mostly rural towns where they are imprisoned. The
'losers' in this shift are argued to be urban communities of colour. Half of all US
prisoners are African American and one-sixth are Latino, with the vast majority
originating from poor urban areas such as East New York and South Central
Los Angeles. As a result, these neighbourhoods, which have already sustained
years of economic and social crises and loss, lose out in terms both of political
representation and power as a result of the rural prison boom. US political districts
are based on population size, and they determine the number of Congressional,
state and local representatives. When political boundaries are redrawn to conform
to census figures, the huge numbers of urban-domiciled prisoners (who are largely
disenfranchised in the US system) are reapportioned to the rural areas where they
are incarcerated. For example, in Florida, the state's growing inmate population

could greatly affect political boundaries in sparsely populated North Florida counties, in which prisons have been constructed in recent years (Huling 2002: 8).

Although Huling's (2002) work emphasises the displacement of people, (and accordingly resources and influence) from urban to rural neighbourhoods, carceral geographers have pushed for a different interpretation of these movements, as connections and linkages, rather than as disjunctures. These rural-urban linkages are exemplified in the phenomenon of 'million-dollar blocks' out of which so many urban residents have been incarcerated, often in rural places, that the costs of their incarceration exceed one million dollars. Building on Gilmore's theorisation of the places from which prisoners come and the places where they are incarcerated as 'a single – though spatially discontinuous – abandoned region' (2007: 31), Mitchelson (2012) has demonstrated, in a quantitative study of imprisonment in the US state of Georgia, a close spatial interdependence between high-incarceration urban communities and prisons located in predominantly rural settings. He identified spatially organised carceral regions anchored by one city or by a cluster of neighbouring cities, with a constellation of prisons in orbit around them. Extending Gilmore's theorisation, he argued that prisons 'are no less important to the urban fabric than are the suburbs, exurbs, and gated communities that similarly 'orbit' large cities' (Mitchelson 2012: 155), and urged that future research should explore qualitatively the particularities of imprisonment in both the cities and their prisons.

Other quantitative work also supports a reconsideration of the connections and similarities between places from which prisoners come, and the places to which they are sent, and nuances the depiction of extra-urban prisons diverting resources from inner-city neighbourhoods to poor rural towns. Engaging with what he terms the central tenet of PIC theory; that the US prison boom was 'supported by prison construction in poor White towns with high unemployment at the expense of Black and Hispanic prisoners' (2010: 1025) Eason used event logistic regression analysis of new prison placements across rural places in the US in the 1990s to show that 'prison boom' prisons were more likely to be built in densely populated towns with an existing prison close by, and with both higher than average percentages of poverty, and of Black and Hispanic residents. On the basis of these findings, he urged that future work reconsider the ways in which race and place inflect and impact prison siting.

The city of Madras, Oregon, provided the setting for just this kind of interrogation of race, place and prison siting, in Bonds' (2013) study of 'YIMBY' (Yes-in-My-Backyard) prison politics. Although lacking the pre-existing prison, Madras otherwise resembled Eason's 'prison boom' towns, with its underlying poverty, and ethnic diversity. Bonds (2013) argued that the YIMBY prison development initiative that brought a 1,800 bed correctional facility to Madras maintained and legitimated local geographies of racialized poverty, inequality and dispossession. The entrepreneurial prison development served to 'simultaneously institutionalise and normalise carceral expansion' whilst sustaining 'local race and class positions' (ibid. 1390). Although Bonds' specific empirical focus was on this rural town, her work is closely articulated with more expansive discourses of racialized mass incarceration and the punitive warehousing of the socially and economically marginalised. As she noted (ibid. 1392–3), rural prison sitings like the one in Madras emphasise exactly the kind of rural-urban connections between 'the carceral expropriation of poor urban communities of colour and the [rural] entrepreneurial recruitment of correctional facilities' highlighted by both Eason (2010) and Mitchelson (2012). Prison recruitment to Madras served to reinforce local racialized (dis)advantage, whilst also underpinning the urban racialized disadvantage coupled to mass incarceration.

Carceral geography has come relatively late to the discussion of prison siting, its advantages and disadvantages, and the complex relationships which start to develop between prisons and their host communities even before ground is broken for their construction. What it brings, though, is an understanding of spatial processes and flows between as well as within locales, and, as Mitchelson (2012: 148) puts it, a desire to 'broaden the conceptual and analytical contexts in which imprisonment is situated'. The perspectives brought by carceral geographers stress the linkages between people and places, the importance of context and the specificity of place in understanding the placement and role of prison facilities, and the wider geographies of marginalisation and opportunity with which they are intimately connected. Carceral geography also brings an understanding of the spatial agility and legacy of carceral systems, and the ways in which the geographies of carceral archipelagos inflect the lives not only of those who live and work in them, but more distant communities to which they are less visibly, but no less viscerally, connected.

Siting, Distance and Pains of Imprisonment

The connections between prisons and communities at distance from them are central to understanding the significance of distance, and how it is experienced. The foregoing sections of this chapter have surveyed the growing literature pertaining to prison siting and the relationships between prisons and their host and source communities. Although the resulting distance between prisoners and their homes is implicit in these debates, its significance tends to be subsumed beneath discussion either of impacts on host communities, or debates over resource distribution. In this section, the direct experience of distance on the part of prisoners, spatially detached from their home places, takes centre stage.

The intuitive sense that distance 'matters' for prisoners, in that the distance from home at which they are incarcerated must have some significance for their experience of imprisonment, has been borne out in a variety of penal contexts – although, as we will see, the interpretation of this significance varies. Conceptualisation of distance on the part of the individual is key, and 'natural elements', such as the physical environment, including climate and the mitigation of its extremes, play an important role. Historically, the challenges of transporting prisoners long distances to new environments have been recognised as practical problems to be overcome, pertaining to the health and wellbeing of prisoners exposed to new and unfamiliar climates and landscapes, with their attendant risks of local pests and diseases, for which new arrivals were unprepared. Climate, seasonality, and landscape, especially when these are new and unfamiliar, all matter to the individual perceiving distance and performing it into being. Shanks et al. (2008) pointed out the physical challenges associated with transporting prisoners from Europe to the Andaman Islands in the nineteenth century, including very high mortality rates due to local strains of malaria, and Wilson and Reid (1949) reported more than half of a group of Allied Prisoners of War perishing from malaria while acting as forced labour for the Siam-Burma Railway. The impact of climate and local physical conditions on prisoners is also well known from classic Gulag memoirs from the former Soviet Union. In *The Gulag Archipelago* Solzhenitsyn (1974: 575–6) described prisoners from the south of the Soviet Union arriving in Arctic Russia in February 1938:

> The railroad cars were opened up at night. Bonfires were lit alongside the train and disembarkation took place by their light; then a count-off, forming up and a count-off again. The temperature was 32 degrees below zero centigrade. The prisoners' transport train had come from the Donbas, and all the prisoners had been arrested back in the summer and were wearing low shoes, Oxfords, even sandals. They tried to warm themselves at the fires, but the guards chased them away: that's not what the fires were there for; they were there to give light. Fingers grew numb almost instantly. The snow filled the thin shoes and didn't even melt ... the doomed prisoners in their summer clothes marched through the deep snow on a totally untraveled road somewhere into the dark taiga. The northern lights gleamed ... The fir trees crackled in the frost.

The distance between 'home' and the prison is not just about the number of miles between two places – it is also about the separation between those places as it is actually experienced by those making the journey; the perceived differences in socially constructed phenomena such as cultural practices, and language, as well as climatic conditions. Among human and social geographers, distance has 'long been a primary target in the struggle against geographical determinism and absolute definitions of space', and theorists of late modernity, postmodernity, and globalization 'have written profusely on the annihilation of space ... by time' (Young 2006: 254). It is argued that distance 'ought not be considered merely as the geographic tract that separates locales, but rather as an active combination of natural, technological, and social elements' (ibid.). In other words, distance should be conceptualised along three dimensions: natural or physical attributes, technological infrastructures that penetrate and/or manipulate spaces, and social relationships among persons in these spaces. In so doing, the 'realism' of distance is multiplied, 'in that the potential configurations of natural, technological, and social elements are exponentially expanded' (ibid. 254), and that the discrepancy in powers to actively configure distances becomes heightened.

The feeling of distance is therefore manifest in a multitude of dimensions. For some prisoners, distance from home is felt less as distance as measured in miles or travel time, and more as distance from familiar landscapes and environments. The US state of Hawaii, facing severe overcrowding of its island prisons, began in the 1990s to 'banish' prisoners to the US mainland, initially to facilities in Texas and later to Mississippi, Oklahoma, Minnesota, Kentucky and Arizona. For these 'exported' prisoners, many of whom had never previously left the islands, Brown and Marusek (2014: 229–230) argued that they experienced these new locations, including 'Appalachian coal country and South-western deserts', as 'alien places', whose distance from home not only foreclosed visitation from family and friends, but more significantly, disrupted 'connections that are vital to a sense of identity, community, and sense of local belonging'. Although this disruption is, they argued, particularly significant in relation to the Native Hawaiian concept of 'Āina [land], this disconnect arguably has resonance for displaced prisoners more generally.

In a very different carceral context, Moran et al. (2011) discussed the experience of distance for prisoners in the Russian Federation. Russia's carceral geography was inherited from its predecessor the Soviet Union in 1991, and that spatial configuration reflected the decisions about prison siting that were made in the context of Soviet ideologies, and in accordance with the needs of the Soviet state. Russia's contemporary carceral geography is characterised by a pattern of 'dispersed concentration' of penal colonies, a throwback from the Soviet era, in which penal institutions are to be found throughout the country, but with concentrations of facilities in some of the more sparsely populated regions of northern European Russia and the Urals. As discussed elsewhere (e.g. Pallot et al. 2010), this geography has bequeathed a system of 'import' and 'export' regions to and from which prisoners are sent in response to the greater availability of bedspace in areas which have functioned as centres of incarceration since the inception of the Soviet Gulag system in the 1930s. For residents of these areas, employment in prison work over generations has engendered a certain sense of patriotism for the penal system which now acts as a disincentive to relocate prison facilities to the regions from which prisoners are 'exported', and which would enable them to be incarcerated closer to home. The mismatch between crime and imprisonment manifest in the spatial 'legacy' of the Soviet era now significantly constrains the extent to which the contemporary Russian authorities could fundamentally reform the distribution of the present carceral estate, should they be inclined to do so. With prisoners dispatched to distant facilities, maintaining close relationships and pro-social bonds, and easing re-entry following release, become intransigent problems which are difficult to overcome within the extant distribution of prison facilities. For prisoners themselves, this distance from home is isolating and marginalising, as the friction of distance affects the ability of friends and family to sustain visiting and other forms of contact. As noted previously (Piacentini et al. 2009: 534) in a study of a facility for juvenile girls in Russia, there was a negative correlation between the frequency of visits and distance from home, in that the greater the distance from home, the less likely it was that prisoners would be visited, and as distance from home increased, the percentage of prisoners who were visited declined. Of course many prisoners in different carceral contexts receive no visits irrespective of how close to home and family they are imprisoned, and for many more the frequency of visits declines as their sentence progresses.

Whereas the Russian study emphasised the negative consequences of distant imprisonment for women and their children and families, suggesting that distance from home may be considered a 'pain' of imprisonment after Sykes' (1959) thesis (Pallot and Piacentini with Moran 2012), and arguing for Russia's spatial mismatch between crime and punishment to be addressed, there are starkly different perspectives on this issue. Research carried out by Bedard and Helland (2004), on women incarcerated in the US, examined similar impacts of reduced visitation associated with increased distance from home, and made a very similar argument, i.e. that distance is a 'pain' of imprisonment. However, whilst mindful of the welfare implications of women's incarceration at distance from home and family, they explored the potential 'benefits' of this situation for crime control. Arguing for a deterrent effect of incarceration, they contended that reduced visitation associated with distance from home increased the punitiveness of imprisonment – in effect making time 'harder' – and they found that increasing the severity of punishment in this way served to increase its deterrent effect. In other words, reduced visitation linked to increased distance from home was shown to deter criminal behaviour. Based on their findings, they debated the cost efficiency of an intentional policy of remote prison building in the United States, considering whether or not the higher transportation and operational costs involved would exceed the increased costs of policing otherwise required to address higher crime rates.

Conclusion and Directions

This chapter set out to provide an overview of the spatial or distributional geographies of carceral systems, and the significance of these geographies for the communities which surround these places, for the communities from which prisoners come, and for the incarcerated themselves. In so doing it has charted a course through scholarship on NIMBY responses to prison siting, and a shift in opinion towards viewing prisons as a stimulus for economic growth, in relation to prisons and their relations with host and 'source' communities. It has also considered the experience of those imprisoned in a carceral estate which for reasons pertaining to its own (past and present) logic of prison siting has placed them at distance from home.

These questions remain of key importance in current carceral systems; for example, in the UK, the National Offender Management Service (NOMS) has engaged in a cost-driven process of prison closure, introducing a new 'estate strategy' in 2010, and closing numerous prisons in subsequent years (National Audit

Office 2013). The resulting geography of UK prisons is arguably characterised by a mismatch of facilities and demand for prison places, for example with the South West of England having more prison places than it 'needs' to serve the local population. This means that spare capacity in some of the isolated, rurally-located prisons in the South West is filled by prisoners from London and the South East, dislocating prisoners from their home communities, making travel for visitors difficult and expensive, and compromising the ability of the prisons to deliver rehabilitative outcomes. Against this backdrop, the UK 'think-tank' *Policy Exchange* put forward a model for radical reform of the UK prison estate, based on a system of new, large 'hub' prisons which they argued would address this spatial mismatch (Lockyer 2013). One of the assumed merits of this approach, a key 'benefit' of the hub prisons, was argued to be their role as economic growth poles; despite the recent evidence from the US, detailed above, of the negligible economic benefit of prison location, the notion of 'profit from punishment' is still driving prison siting policy.

In considering next how carceral geographers can advance this key field of inquiry, the following two chapters identify and develop themes which emerge from this work, and which are implicit in the operation of carceral systems, whether for the communities which host them, or for the prisoners who inhabit them. The following two chapters in this section accordingly focus first on mobility, as an inherent part of the functionality of carceral systems, particularly significant where prisoners must travel to serve their sentences, and next on the nature of the boundary between what is considered 'inside' and 'outside' of the prison, turning a spotlight on these two aspects of carceral systems in ways which resonate with contemporary themes within critical human geography.

Chapter 7
Prison Transport and Disciplined Mobility

There is a significant weakness in the literature on mobility – namely the undertheorisation of mobility and power (Moran et al. 2012). The 'mobilities turn' has tended to draw a connection between mobility, autonomy and freedom, and in so doing has inadequately explored and theorised coerced mobility. However, recent scholarship of the contemporary spaces of incarceration and carceral practice offers a perspective on, and empirical examples of, forced, coerced, punitive, disciplined or governmental mobility, and while carceral geography has increasingly focused attention on the mobilities inherent in carceral practices, there is scope to pursue this line of enquiry in greater depth.

It is understandable that the first geographical studies of incarceration tended to overlook mobility in the carceral context. Prison seems inherently spatially 'fixed', and prisoners in turn immobile by virtue of their imprisonment. However, criminologists have long been aware that mobility is a constant practical concern in the management of penal systems, with, for example, prisoner transport has been identified as an issue in relation to overcrowding (Wooldredge 1991), a risk environment for the transfer of infectious diseases (Levy et al. 2003), for sexual coercion (Hensley et al. 2003), and for sexual taunts and verbal abuse between transported prisoners (Scraton Moore 2005). There have also been concerns to reduce the costs and inconvenience of inmate movement in connection with healthcare provision (Stoller 2003), for example by providing intermediate mental health care (MacKain and Messer 2004) and 'telemedicine' (Zollo et al. 1999). Within carceral geography, the work of Martin and Mitchelson (2009: 461) was amongst the first to draw attention to the significance of the mobility of both prisoners and guards within and between penal institutions.

This chapter provides an overview of the treatment of mobility within carceral geography, in the light of the recent explosion of literature on mobility in contemporary human geography (Moran et al. 2012). The 'mobilities turn' tended towards a conflation of mobility, autonomy and freedom, and in so doing, left coerced mobility underexplored and undertheorised. This chapter argues that geographical studies of contemporary spaces of incarceration and carceral practice have begun to offer a perspective on and an empirical example of what might be termed disciplined, forced, coerced, or governmental mobility. The chapter suggests that the notion of the 'carceral' is reflexively and recursively useful not just for studies of incarceration *per se*, but also for understanding the restriction of autonomy in a much broader context.

The chapter first provides a brief overview of the 'mobilities turn' as it took form within contemporary human geography, and a discussion of the ways in which this growing body of literature engages, or does not engage, with notions of forced or coerced movement. There then follows a discussion of recent scholarship within carceral geography, which addresses such forms of movement head on, using examples from prisoner transport and the movement of asylum seekers, and the issues faced by prisoners in relation to restricted movement within confinement. The chapter then moves to a more detailed discussion of specific studies of mobility within, between, and beyond carceral institutions.

Mobility and Power

Described as an 'evocative keyword' (Hannam et al. 2006: 1) for the twenty-first century, a wide-ranging discourse has developed around mobility, covering the actual or virtual movements of people through space (Sheller and Urry 2006, Silvey 2004, Urry 2007). Urry (2002) posed the fundamental question, 'why travel?', but despite the breadth and diversity of responses, one sphere of mobility has remained under researched. Implicit or explicit in the majority of both the empirical and conceptual literatures is a sense that mobility is connected with autonomy, the realisation of potentials, and ultimately, 'freedom' (Moran et al. 2012). Dunn (1998: 43) has described mobility as 'the potential for movement', Knie (1997: 143) referred to the 'construction of possibilities for movement', and Sörenson (1999: 3) highlighted the 'performance' of mobility, all to a greater or lesser extent presuming that the individual has the agency to determine their own mobility. Observing that these definitions and stances on mobility reveal two 'contrasting facets of mobilities', Uteng (2009: 1057) identified on the one hand the positive associations with progress and freedom (Sager 2006), and on the other the 'issues of restricted movement, vigilance and control'. Hannam et al. (2006: 3), drawing upon Massey (1994) described mobilities as 'caught up in the power geometries of everyday life'. The focus here is on power, and [access to] mobility, and as Skeggs argued (2004: 49), '[m]obility and control over mobility both reflect and reinforce power. Mobility is a resource to which not everyone has an equal relationship'.

Despite these nuanced conceptualisations of mobility as reflecting and reinforcing power, in line with the seemingly innate connection between mobility, rights and autonomy (Cresswell 1999, 2006), the focus of early empirical work on mobility tended to be on *access to*, and *exclusion from* it, as manifestations of power (see, for example Uteng 2009 on constrained mobility in Norway, Freudendal-Pedersen 2009 on urban mobility in Copenhagen, and Richardson and Jensen 2008 on the Bangkok Sky Train) in which mobility is understood as a specific kind of ontological object, rather than as a quality or characteristic. Until recently, far less attention had been paid to another type of relationship between mobility and power, in which mobility is an *instrument* of power. 'Involuntary' mobility was acknowledged and labelled; such as by Hannam et al. (2006: 10) as 'obligatory' travel ('necessary for social life, enabling complex connections to be made, often as a matter of social or political obligation'), or 'coerced' movement, such as that of refugees, or of forced migrants (Indra 1998, Kofman 2002). The movement of persons who are mobile, but not 'through their own volition' was also recognised, such as in the work of Cloke et al. (2003) on the homeless in the UK, whose lives are unsettled and constantly 'on the move', and of Wolch et al. (1993) on the 'Greyhound therapy' of US states supplying the homeless with bus tickets to cross state lines. With these few notable exceptions, these early literatures on mobility had a tendency to elide the involuntary, obligatory or coerced nature of mobility, with the result that consideration of coercion – what might be described as forced or 'disciplined' mobility – was missing from the mobilities discourse. This omission was perhaps because 'coerced' movements, which could for example include trafficking, extraordinary rendition or kidnap, had not yet been adequately theorised within the 'new mobilities paradigm' (Hannam et al. 2006, Sheller and Urry 2006), even during the emergence of a 'politics of mobility' (Cresswell 2010).

Previous work has argued that connection between mobility and autonomy as a logic of power needed to be looked at closely, and suggested that one lens through which it may be examined is Foucauldian Panopticism and biopower (Moran et al. 2012). As Packer (2003: 140) has discussed, mobility can be described as 'one of the material corollaries' of autonomy, but he also argued that mobility and autonomy must be understood 'in their specificity and in their necessity to current forms of governing', in that 'as the potential for mobility is increased, the subject of governing must change in accordance', it is possible to envisage governing 'at a distance' taking the place of 'direct control' (ibid.). Such 'disciplined mobilisation', in which mobility is a tool of governmentality, is necessary for 'many of the social and economic changes said to characterise neoliberal capitalist societies' (2003: 143). Taking self-regulation to be the ultimate form of Foucauldian Panopticism, one could argue that mobility is never genuinely 'free' or autonomous in as much as it takes place within a system evolved to deliver the *types* of freedom which satisfy the demands of culture and economy, and in which citizens are 'carefully fabricated, according to a whole technique of forces and bodies' (Foucault 1979: 217). By considering mobility and autonomy in this way, a space can be opened for the consideration of 'disciplined mobility', bringing the issue of coercion more clearly into view.

The nature of autonomy is itself problematic, and in some readings, such as that of Bevir (1999: 66) it appears to be absent from Foucault's theorizations; the subject seemingly rendered impotent in the face of immanent and all-pervasive power, since it only 'comes into being as a construct of a regime of power'. Others argue that autonomy exists *within a realm of possibilities* defined by discursive practices' (Wisnewski 2000: 434, my emphasis). In this interpretation, '[t]he fact that we are always in relationships of power does not entail that we cannot make choices freely … ' (ibid.). Foucault seems to allow room for a '"weak" autonomy, understood as the ability to make more or less free choices within a realm of possibilities' (Wisnewski 2000: 435), a situation described by Bevir (1999) as a distinction between autonomy and agency. As discussed in Chapter 3, for Bevir, Foucault's rejection of autonomy does not necessarily entail the rejection of agency. Since different people respond differently to the same social structures, there must, he argues, be an 'undecided space in front of these structures where individuals decide what beliefs to hold and what actions to perform' (1999: 68) effectively making choices within Wisnewski's realm of possibilities. The subject is, therefore, an agent, even if not an autonomous one (Bevir 1999).

This notion of agency within a limited range of possibilities characterises the different kinds of subjects produced in Foucault's theory of discipline, and it has direct relevance to discussion of the nature of agency within mobility. The agency evident in boarding a commuter train differs wildly from that deployed in extraordinary rendition (Moran et al. 2012). Previous work discussed and advanced the notion of 'disciplined mobility', and argued that it offers a means of conceptualising mobility with limited agency, suggesting that geographers might envisage a spectrum along which to locate mobilities – commuting, air travel, human trafficking, extraordinary rendition, refugee status, asylum-seeking, and prisoner transport, which all share the characteristic of 'disciplined' mobility, differentiated by the production of subjects with a greater or lesser degree of coercion or agency (Moran et al. 2012).

Mobility in Carceral Geography

Prison *appears* inherently spatially 'fixed', and prisoners in turn *seem* to be immobile by virtue of their imprisonment (Moran et al. 2012). Although the mobilities turn arguably overlooked disciplined mobility, carceral geography has focused closely on the mobilities inherent in confinement, and has recently begun to emphasize the significance of circulations of ideas, things, practices, around spheres of confinement (Gill et al. forthcoming). Spatial regulation is a key form of social control (Foucault 2003), and theorists have identified two predominant means through which spatial control is exercised; expulsion and containment (Beckett and Herbert 2010). As a form of containment, prisons may seem to be the epitome of immobility, with inmates incarcerated within a static physical space. However, the rise of the penitentiary partially overshadows previous forms of punishment for offenders; including expulsion, or the forced removal of specified individuals or groups from a designated territory. Although transportation is now considered something of a historical curiosity, the organized transport of convicts as a form of spatial regulation was widely employed, in Babylon, Greece and Rome, and was the principal form of punishment in Britain in the eighteenth century. The British state relied heavily on transportation to the American and Australian colonies for 150 years (Beckett and Herbert 2010, Willis 2005, Casella 2005). Foucauldian analysis suggests that convict transport and the reformatory/penitentiary are qualitatively different and temporally discrete modes of punishment, in that the rise of the penitentiary saw the decline of transportation, but this distinction has since been challenged by more contemporary accounts (Moran et al. 2012, Pallot 2007, Willis 2005).

With the recent 'mobilities turn', geographers are ideally placed to interrogate mobility in the penal context, and accordingly, a distinct theme of work has emerged within carceral geography. As noted in previous chapters, geographers have focused on the nature of carceral spaces and experiences inside of them (e.g. Baer 2005, Gear 2005, 2007, Dirsuweit 1999, Sibley and van Hoven 2009, Moran et al. 2009) and the spatiality of the relationship between the emplaced correctional institution and the wider community (e.g. Baer and Ravneberg 2008, Peck 2003, Peck and Theodore 2008, Gilmore 1999, 2002 and 2007). Within this literature, they have considered mobility in terms of the location of the physical prison itself, through study of the decision-making process and its influence on local communities (Bonds 2006, Farrigan and Glasmeier undated, Burayidi and Coulibaly 2009), or of the micro-scale mobility of inmates and staff within a penal institution (Philo 2001). As Martin and Mitchelson (2009: 461) observed, the mobility of both prisoners and guards within and between penal institutions means that 'contemporary practices of imprisonment are characterized by [the] tensions between apparent fixity and forced mobility'.

Thinking of penal institutions as islands in a carceral archipelago, authors are increasingly considering the nature of mobilities *within*, *to* and *from*, *between* and *beyond* these islands. Accordingly, this chapter now turns to an overview of research into these mobilities.

Within Institutions

Considering first mobility within institutions, prisons can be understood as the intentional translation of power relations into the organisation of space and movement for the purposes of production (Massey 1995), where production includes both the production of prisoners (after Foucault) and the economic production of prison staff wages and prisoner labour. In respect of prisoners' ability to physically move around and access resources and facilities, mobility within institutions is closely connected to a limited degree of autonomy, the realisation of potentials, and ultimately, the nature of 'freedom' within incarceration. It resonates with Dunn's (1998: 43) description of mobility as 'the potential for movement', Knie's (1997: 143) reference to the 'construction of possibilities for movement', and Sörenson's (1999: 3) highlighting of the 'performance' of mobility, all predicated to a greater or lesser extent upon the ability of the individual to determine their own mobility. Of course these definitions reveal two 'contrasting facets of mobilities' (Uteng 2009: 1057); on the one hand mobility's positive associations with progress and freedom (Sager 2006), and on the other the 'issues of restricted movement, vigilance and control'. Hannam et al. (2006: 3), have described mobilities as entwined with the power geometries of everyday life. As Hyndman (2012: 248) has argued, the mobility of all persons 'is subject to the calculus of Massey's (1993) power-geometry, but conditions of highly restricted mobility, even containment, are more common for those bodies that are criminalised, displaced, and/or construed as a security threat to the state and its citizenry' (Cresswell 2006).

In relation to the power geometries of prisoners' everyday lives, in different facilities, different prisoners serving sentences under different regimes, are afforded different levels of mobility, and according to their personal circumstances (including any physical impairments), are differentially able to take advantage of opportunities for movement. Prisons are, by their very design, nested pods of space to which prisoners, guards, other prison staff, prison visitors and so on have different extents and levels of access, at different times, and under different circumstances. Layered on top of this inherent and physically engineered, securitised and managed restriction of local mobility within the physical space of the carceral institution, exists a further level of (usually unintended, and recognised as problematic) restriction on mobility imposed by physical impairment. The wider mobilities literature within human geography is in dialogue with discourses in disability geography, in which a rich body of work has emerged on the materiality of the impaired body and its experience in 'ableist' or non-disabled spaces (e.g. Hansen and Philo 2007, Hall 2000). Both the 'medical model' of disability studies, focusing on the 'damaged' body or mind which requires self-adjustment to function in ableist space, and the 'social model' (far more widely adopted within human geography), which argues that it is the wider society that fails to accommodate bodily impairment, have relevance for the experience of physically or functionally impaired prisoners. Although prison systems are obliged to make allowances, and to provide facilities, for physically (and mentally) impaired prisoners, the extent to which these usefully mitigate against the restrictions on mobility within the institution varies widely.

Prisons are overwhelmingly 'ableist' spaces, with living accommodation poorly adapted to occupation by wheelchair users, for example, and provision of care personnel often inadequate to cater to the needs of physically impaired prisoners, who frequently rely on cellmates, or other prisoners, to assist them in their daily movements. In their study of older women prisoners in the US, Williams et al. (2006) found that functional impairments coupled with the prison environment led to negative outcomes for a significant proportion of their respondents. Specific 'prison environmental stressors' exacerbated these women's situations, with 29 per cent of the older women surveyed assigned to top bunks (in a setting where swapping bunks is prohibited) into which they found it very difficult to climb. Williams et al. (2006: 5) noted that although there was a mismatch between functional impairment and living environments in the community in general, this was intensified in prisons, in two significant ways; firstly, because prisons tend to be designed for young, healthy inmates without functional limitations, and secondly because the prison regime requires certain activities such as occupying top bunks, and dropping to the floor in response to security alarms, which can be challenging for physically impaired inmates. The problems of physically impaired, and in particular, ageing, inmates, are beginning to take prominence in prison research, largely in response to a realisation that prison populations are themselves ageing; for example the US prison population is ageing at a significantly more rapid rate than the overall population, with the population of older adults in prison having more than tripled since 1990 (Williams et al. 2012). However, more research is needed to guide the implementation of appropriate care models in carceral settings.

Although fellow prisoners may be willing to assist physically impaired prisoners, and in some contexts, may receive payment or other consideration for the help provided, it is often observed that the nature of the crimes for which physically impaired prisoners have been incarcerated renders others unwilling to assist them. Touraut's compelling work[1] on the proximity of mobility and liberty in prisons in France investigated the experience of carceral space on the part of older prisoners, who as they age and become less physically mobile, not only encounter limited mobility within prison spaces ill-designed for their needs, but also face prejudice based on assumptions (rightly or wrongly) that they are sex offenders. Testimony from her interviewees brought vividly to life the marginalisation suffered by these inmates, some of whom are restricted to living on one floor of their wing of the prison, never accessing outside space, because of the lack of lifts between floors, or who rely on being carried down staircases, or manoeuvred around their cells, by fellow inmates. Personnel in some of these prisons refused to assist prisoners in these ways as this kind of caring work falls outside the remit of their job description, and often the level of assistance provided by dedicated care staff is insufficient for prisoners' needs.

Where direct and targeted care *is* provided to prisoners with severe physical impairment, medical staff may encounter challenges in delivering this care within the carceral environment, and in supporting the kind of independent mobility which is commonly encouraged in patients convalescing in civilian settings. For example, in a rare empirical study of the recuperation of a paraplegic prisoner, paralysed after a fall and returning to prison in the US after medical care in hospital, nurses delivering rehabilitative care found direct practical contradictions between the independence, mobility and self-care which they would usually advocate, and the barriers of confinement, compliance, and physical restrictions imposed in prison, as well as having to overcome their own reticence about working with a prisoner and anxieties about providing intimate care in the presence of the prison guards who accompanied the prisoner in hospital (Graham and Royster 1990, Benoliel 1990).

In her work on access to prison healthcare for women in the United States, Stoller (2003) described mobility in terms of the 'nested' nature of prison spaces (the bunk, inside the cell, inside the dormitory, inside the unit, inside the prison), the directionality and permeability of boundaries between prison places, and the

---

1   Presented at the Terrferme colloquium 'Confinement viewed through the prism of the social sciences: Contrasting facilities, confronting approaches', University of Bordeaux, October 2013.

differential permeability and accessibility of spaces to different types of individuals such as prisoners, healthcare workers, and prison guards. In prison settings, she observed, 'prisoners have no option but to conform in their movements, their arrangements of meagre possessions … and their daily schedules, even within their cells, to the rules of the larger institutions' (Stoller 2003: 2265). Further,

> the directional permeability of the cell, clinic and exit doors vary with the social role of the institutional inhabitant. Prisoners can enter or leave their cells only at prescribed times. Depending on the institution and the time of day, they may require written permission to leave their units … although medical staff can come and go to their clinics, they cannot enter a prisoner's cell without a guard accompanying them. Guards as a class can generally go anywhere in the institution, although individually they are assigned to work in specified areas. (ibid. 2265)

She described the case of Luisa Montalvo, a prisoner who died having failed to access healthcare due to various spatial practices of imprisonment; which forced her to lie on the floor in an overheated cell (because prisoners were not allowed to swap bunk beds and cell doors had to be locked closed even in hot weather), required her to walk to the 'pill line' to collect medications in person, and meant that medical staff were unable to enter her cell to treat her without the presence of a guard. No wheelchair was available to her since, having not been assessed by medical staff, she had not officially been ruled 'in need', and throughout her illness she was under pressure to return to her prison job of cleaning floors, and struggled to get medically-approved 'lay-ins' to avoid being disciplined for missing work.

In carceral settings, then, mobility is a key facet of the power geometries of prisoners' everyday lives. Autonomous mobility is heavily restricted; prisoners are subject to flow control, to daily regimes of movement, and are barred from entering and exiting certain areas of the prison. Where a form of autonomous mobility is allowed, for example in movement around recreational spaces within the prison, or within living accommodation, further restrictions are imposed by functional impairment and the lack of fit between the impairment and the facilities of the prison environment – the meshing of the 'medical' and 'social' models of disability studies discussed earlier. For some prisoners, imprisonment is not only confinement within the physical institution, but also within their physical bodies, for which the prison setting may be very poorly adapted, and for whom mobility is further restricted.

To and Between Institutions

Mobility to, from and between facilities is a key issue in the management of penal systems, but until relatively recently, research has tended to focus on the cost, efficiency, security and practicality of movement (Wooldredge 1991, Hensley et al. 2003, Levy et al. 2003, Scraton and Moore 2005, Stoller 2003, MacKain and Messer 2004, Zollo et al. 1999) rather than on how it is personally experienced. However, recent research into prison transportation and movement to and between institutions, stresses its social, personal, emotional and transformative effects, in the context of the power relations of imprisonment.

Although studies of prisoner transportation are relatively few, a common theme in all of them is of the dehumanising and humiliating effect of movement for the transported. In their study of the social consequences of prisoner movement in Sweden, Svensson and Svensson (2006) noted that moving prisoners enabled the authority of the mover to be performed and enacted and to reinforce the identity of the prisoner as a subjugated body. They also noted that transportation had a transformative effect that was often intentional, in that prisoners were frequently moved in response to their unruliness, or in order to disrupt 'cultural rules and rituals … establishing amongst the prison population' (Svensson and Svensson 2006: 2); and that the act of transportation was itself intended to effect in them some change in behaviour, which necessitated the constant enforcement of power relations during the transportation itself.

Of course it is not only convicted prisoners who are moved between institutions – as several scholars have noted, migrant detainees also suffer the negative effects of enforced mobility as they move between holding facilities. For example, Gill (2009) noted in his study of the movement of detained asylum seekers around the detention estate in the UK, that movement, and even just the threat of movement, caused considerable stress, confusion and anxiety to detainees, whose continual movement arguably reduced their external visibility to supporting agencies, in a process termed 'ghosting' (Wilson 2008, in Gill 2009). The continual movement of detainees, he argued, serves to position them as 'transitory, fleeting and depersonalised to those actors with the greatest influence over them' (Gill 2009: 186), with profound implications for their welfare and their detention outcomes.

Hiemstra (2013) mapped the detention paths of migrants detained in US facilities, and described a chaotic system of movement in which the average number of transfers amongst her interviewees was 3.4, with 10 detained in more than four different places across the United States. One detainee was held in a total of six different locations over 12 weeks, having been detained in New Jersey, then transferred to Manhattan, then to Pennsylvania, before moving on to one of two different locations in Texas, and finally to Louisiana, from where he was eventually deported. She found tracing the paths of other detainees more challenging because the extent of their disorientation was such that they simply did not know the names or locations of the places to which they had been moved. Detainees in her study complained of the traumatic conditions of transfer, during which they were routinely shackled, hands and feet to waist, had to wait in chains for long periods whilst being checked out of one facility and into another, and described cold, crowded rooms without space to sit or lie down and insufficient food being provided during transport. In addition, detainees described transportation as disorienting and upsetting, frequently roused in the night to be moved with little or no notice, to places whose names and locations they did not know, and with a sense of secrecy that meant that even when they were being moved internally within the United States, many feared that they were actually being deported out of the country. Repeated and extended transportation also served to complicate detainees' paperwork trail, with information and documents lost or delayed, and new procedures required at different facilities. Essentially, Hiemstra (2013: 70) argued, 'the cumulative architecture of the detention and deportation system makes detained migrants feel like they are in a purgatory of waiting, a feeling with profound implications'. Some become so tired and disheartened by the 'Kafkaesque' (ibid. 71) process that they sign their deportation papers in order simply to escape the process.

These studies from three different locations all have much in common with previous work (Moran et al. 2012) conducted with Russian women prisoners experiencing the Russian system of prison transport, called *etapirovaniya*, or '*etap*'. Although *etap* is an example of carceral practice from what has been argued to be an exceptional carceral system, many of its characteristics have considerable resonance with prisoner movement systems in other carceral contexts.

In the Russian system, persons given custodial sentences are allocated to a prison 'colony' in which their sentence will be served. The colony is chosen by central penal authorities according to factors including the Russian Penal Code, the duration of the sentence, the appropriate regime of facility (for men), and the space available in appropriate colonies. The process of transportation is organised centrally and usually takes place by train, in specially converted wagons divided up into cells, frequently without washing or toilet facilities. Prisoners often lack information about the nature and the timing of their departure for the colony, causing disruption in communication with friends and family. Many are not told the colony to which they are travelling, discovering this only during *etap*, or on arrival at the colony itself – as one respondent[2] noted

> In the first place, they don't tell you where they are sending you ... We were in the dark about it, and when they put us on the *etap*, they didn't mention it. It's not talked about. I was told that it's a secret. You get ready, and you go on *etap*, and it's all kept a secret. (Moran et al. 2012: 452)

Even if told their destination, prisoners have no control over, and often have no knowledge of the likely duration of the journey, which can be unpredictably lengthy and circuitous. Given the scale of the Russian Federation, the distances covered can be vast, and tales of weeks or months in transit are common. Previous work discussed the examples of two women convicted in the same Russian city, and sentenced to the same prison colony, whose *etap* journeys were long both in distance and duration, but varied enormously (Moran et al. 2012). Whereas one

---

2   Interviewed for research conducted with Judith Pallot and Laura Piacentini, 'Women in the Russian Penal System'. ESRC award RES-062-23-0026.

was in transit for less than a month, travelled relatively directly, with three stops en route, the other was in transit for almost three months, and took an entirely different route, frequently doubling back, making several additional stops, and covering hundreds of extra kilometres.

The timing, duration and route of the *etap* journey are all outside of prisoners' control, and the conditions on *etap* served to further emphasize the disciplined nature of this mobility (Moran et al. 2012). Kept in cellular confinement in frequently overcrowded, often windowless train wagons, under direct surveillance by armed guards with dogs, women prisoners were corporeally restricted in terms of nutrition, and toilet and washing facilities.

The experience of women transported within Russia's prison system has much in common with political prisoners moved between prisons in former East Germany (GDR), who experienced a transport system inherited from Soviet post-War control of the Occupied Zone. Published testimony from prisoners moved in prison trains in the 1960s and 1970s suggests that prisoners were not told their destinations, guessed at their whereabouts *en route* from loudspeaker announcements on train station platforms as they passed through; that frosted and barred windows prevented any view of the outside world, that prisoners were handcuffed throughout their journeys, and that conditions inside the trains were cramped and restrictive. A rail coach used for these transportations is preserved at the Gedenkstätte Berlin-Hohenschönhausen, 'Stasi Prison' Museum in former East Berlin (see Figure 7.1). Long journeys to prison were due to trains following a predetermined loop-shaped route, and travelling by night to avoid congested lines. Harmut Richter, interviewed in 2004, recalled the circumstances of his journey in 1966, when he was transported aged 18 to stand sentence for attempting to leave the GDR:

> My trip with the Grotewohl-Express from Karl-Marx Stadt (Chemnitz) to Potsdam lasted six entire days. "We will shoot you if you try to escape" a guard informed us right at the beginning. There were about fifteen of us prisoners. On the way to the train station, we were handcuffed to one another. The guards had additional control over us with toggle chains. Attendants with dogs accompanied our party. I was crammed into a narrow cell together with three other men. You could neither sit nor stand properly. The lavatory in the train was filthy with excrement and urine, and whoever wanted to use it had to call and knock until one of the guards took pity. For "interim custody" we were put into dirty transit cells of other prisons. I got to know the fates of many people that way, despite the attempts to prevent contact with prisoners in the adjacent cells by using the "socialist persuader", as we called the rubber cudgel. The deserted train stations carried familiar slogans such as "Help plan, help work, help govern!", or "The human being is at the centre". It was on this ride that I lost my belief in a socialism with a human face. (Gedenkstätte Berlin-Hohenschönhausen, Interpretation Panel, August 2012)

**Figure 7.1   Prisoner transport rail coach preserved at the 'Stasi Prison',
Gedenkstätte Berlin-Hohenschönhausen, former East Berlin**

Within the wider mobilities literature, which emerges out of a particular
cultural specificity bearing little resemblance to the circumstances of Russia or the
former GDR, research into the use of travel time tended to detail activities such
as working, the reading of books and newspapers, the use of mobile electronic
equipment, and listening to music as time-filling pursuits (e.g. Lyons et al. 2006).
However, this literature has also usefully considered the 'production' of travel
time, with Urry (2006: 368) describing travel as 'distinct social practices involving
different kinds of experience, performance and communication' and Watts (2008:
711) suggesting that passenger time 'is not a simple flow but a percolation' in
which 'passenger times coalesce … to form communities'. Although testimony
from those who have experienced forced mobility commonly includes descriptions
of the harsh, cramped conditions, prisoners also speak about having to 'get by'
with their fellow passengers, and for some, a sense of camaraderie developing on
the journey. As the last period of time before the confiscation of civilian clothes on
arrival at the colony, for Russian prisoners *etap* is a liminal space of betweenness –
'a mobile penal space of learning and acclimatization to prison life, accessorized
with some of the material trappings of freedom' (Moran et al. 2012: 456, 2013).
For Harmut Richter, the circumstances of transportation, along with contact with

other prisoners and reflection on state slogans, disabused him, and presumably many others, of any lingering vestiges of belief in the prevailing socialist system.

Prison transport is an example of the type of disciplined mobility scarcely considered in the mobilities literature (Moran et al. 2012). Prisoners commonly lack agency in terms of the timing, nature, route, and physical circumstances of travel, vanishing inside 'black-box mobility', beholden to transport personnel. For Russian women prisoners, the experience of *etap* served directly as form of disciplinary power, in which travel time facilitated the acclimatisation of prisoners to the institution of the prison.

The movement of prisoners, whilst deeply significant for them personally, also has important ramifications for penal systems and for the political contexts within which they develop and operate. In divided Cold War Germany, for example, prisoners became a profitable commodity to be traded and moved cross-border for direct economic and political benefit. Political prisoners in East Germany were traded with West Germany in return for hard currency, in a system which partially rectified the chronic shortages of goods in the East German planned economy (Palenberg 1982, Horster 2004). This system of prisoner purchase had by the 1970s settled into a practical procedure whereby East German prisoners were moved internally to Chemnitz prison (Strafvollzugsanstalt Karl-Marx-Stadt) where they were collected by coaches and, having signed away their East German citizenship, were taken over the German border to an Emergency Refugee Camp at Giessen in West Germany. The process was expedient both for East and for West Germany, and for the prisoners themselves, many of whom had been imprisoned for trying to leave East Germany in the first place. Some may even have intentionally pursued arrest and imprisonment in the full knowledge of the likelihood of their sale through the purchase system, and with the intention of recommending other political prisoners for purchase by the West German state. For East Germany, apart from the economic benefits, the purchase scheme operated as a safety valve to release pressure on the prison system, and to weaken protest movements. For West Germany, the purchases were relatively cheap, and any contributions to improving living conditions for German citizens in the East (in a country not formally recognized by West Germany) were electoral vote-winners.

Albeit in a very different context, in their work on the expansion of privatised imprisonment in the United States, Welch and Turner (2007: 62) also drew attention to the transportation of prisoners between facilities as a means of profit generation; 'dynamic incarceration that renders commodified prisoners mobile is good for business'. They found that prisoners in the United States were frequently transferred between facilities in their own state, and increasingly across state lines, and they argued that this cross-state transportation was for the sole purpose of occupying cells in private prisons and detention facilities, in a system which exploits the economic value of prisoners detained at state expense. Whereas in the federal justice system it might be expected that the federal state would dispatch prisoners to distant facilities, they noted that for individual *state* systems to do this was particularly significant. The distances travelled by prisoners transported in this way can be immense – they noted that the Correctional Corporation of America's (CCA) facility in Arizona housed prisoners from Hawaii and Alaska, and that its facility in Tennessee imported prisoners from Vermont.

Although it was earlier noted that despite the emergence of a 'politics of mobility' (Cresswell 2010), 'coerced' mobilities, such as human trafficking or extraordinary rendition, had not yet been adequately researched and theorised within the 'new mobilities paradigm' (Hannam et al. 2006, Sheller and Urry 2006), recently published fragmentary accounts and descriptions of such coerced movements convey a strong sense of the intentional disorientation enabled by multiple movements, quite apart from the human rights abuses enabled by these secretive detention processes. Two examples:

> On February 12, 2003, at around 12:30 pm, Mr Osama Mustafa Nasr (Abu Omar) was walking from his house in Milan to the local mosque. He was stopped by a plain-clothes *carabiniere* (Italian military police officer) who asked for his documents. While he was searching for his refugee passport, he was immobilized and put into a white van by more plain-clothes officers, at least some of whom were US agents. He was brought to a US airbase in Aviano in Northeast Italy and from there, flown via the US airbase in Ramstein, Germany to Egypt. He was held for approximately seven months at the Egyptian military intelligence headquarters in Cairo and was later transferred to Torah prison. His whereabouts were unknown for some time, and he was allegedly tortured. He was released in April 2004, rearrested in May 2004, and eventually subjected to a form of house arrest in Alexandria in February 2007. (Messineo 2009: 1023)

> Shuttling detainees into the facility without being seen was relatively easy. After flying into Bucharest, the detainees were brought to the site in vans. CIA operatives then drove down a side road and entered the compound through a rear gate that led to the actual prison. The detainees could then be unloaded and whisked into the ground floor of the prison and into the basement. The basement consisted of six prefabricated cells, each with a clock and arrow pointing to Mecca, the officials said. The cells were on springs, keeping them slightly off balance and causing disorientation among some detainees. (*Inside Romania's Secret CIA Prison* [December 8, 2011] cited in Carey 2013: 447)

These two extracts, and the work of Perkins and Dodge (2009), drew attention to the extreme coercion involved in these mobilities, through which those renditioned are held against their will as they cross international borders, traversing jurisdictions with the primary intention of locating them in a context in which information can be extracted via extra-legal means. They commonly lack information about their whereabouts, and are intentionally concealed from legal assistance, and from contact from friends and family. Although these practices remain under-researched, the highly controversial process of rendition has already commanded the attention of scholars of borders, migration, asylum and geopolitics, and recent work has drawn attention initially to the mapping of the routes of rendition through numerous sovereign territories, and to the political processes of disclosure which have followed exposure of sites of rendition (e.g. Carey 2013).

Examples such as these also draw attention to the relationship between differential coercion and volition in disciplined mobility more generally; for example in the case of trafficked workers (Brunovskis and Surtees 2008, Kim 2007, Srikanthiah 2007), autonomies in extraordinary rendition (Sidaway 2010), and the agency of asylum seekers (Gill 2009, Schuster 2005, Rygiel 2011, Mountz 2011). Away from these examples of extreme denial of agency, it may also be useful in surfacing the structures of discipline and autonomy underlying the 'everyday' forms of mobility most frequently investigated in the global North, and in uncovering the ways in which the performance of mobility reflects the modes of governmentality within which it takes place.

Conclusion

This chapter has outlined the ways in which carceral geographies have explored the cartography of imprisonment through a focus on the mobilities inherent in carceral practices, both within and between institutions, through work which traces the (restricted) movements of prisoners within institutions, and which maps the coerced mobilisation of captives of carceral systems as they are moved to and between institutions. As Sidaway (2010: 673) noted in relation to extraordinary rendition, though, 'beyond such stark and simple maps are myriad human geographies which the maps are unable to do justice to'. As the work discussed has shown, although tracing and mapping movements is one step towards understanding their significance, just as important is gaining an understanding of the experience of those movements themselves and the power relations inherent in them and reinforced through them.

Adey (2006: 91) cautioned that in order to be taken seriously, the 'mobilities turn' needed to realise the relations and the differences between movements; 'if we explore the mobility in everything and fail to examine the differences and relations between them' there is a danger of mobilising the world into a 'transient, yet featureless, homogeneity'. Carceral geography is actively exploring mobility in the carceral context, and is drawing attention specifically to the differences between movements, thus directly addressing the neglect of coercion in mobility in the wider literature. The empirical example of *etap* described in this chapter has been argued to be an example of coerced movement (Moran et al. 2012) conceptualised as 'disciplined mobility', drawing on Foucauldian understandings of discipline and governmentality, in which mobility is an instrument of power and in which the subject of mobility has limited agency in the process of movement. Although in this work *etap* was posited as an extreme account of disciplined mobility, with the limited agency of the subject having resonance with, for example, trafficking and extraordinary rendition, as the discussion in this chapter has demonstrated, *etap* represents just one position in a spectrum of disciplined mobility, which exists within and between carceral institutions. And more broadly, mobility, although commonly posited as an expression of autonomy and freedom, is *always* disciplined; even in circumstances where mobility *can* be described as the exercise of autonomy, the choice to move occurs '*within a realm of possibilities* defined by discursive practices' (Wisnewski 2000: 435).

The types of mobility inherent within carceral systems by definition involve the traversing of the assumed boundary of an institution – from 'free' life to captivity, between captive spaces, and from captivity to 'freedom'. The following chapter focuses closely on this boundary-crossing, directly addressing the nature of the prison boundary, and the assumed binary opposition of inside and outside.

Chapter 8
Inside/Outside and the Contested Prison Boundary

The notion of the 'carceral' as a social construction existing both within and separate from physical spaces of incarceration is central to this chapter, which draws upon the recent surge in scholarship theorising the nature of the carceral boundary and critiquing Goffman's notion of the 'total institution'. The assumed binary between 'inside' and 'outside' of prisons has been heavily critiqued, and in drawing out the threads of the critical discourses, this chapter draws attention both to the indistinction of the prison boundary itself, and the specific spaces nominally 'within' prison which act as liminal sites of boundary crossing or blurring; and to a wide range of diverse sites and circumstances 'outside' prison which are either characterised by the replication of aspects of incarceration, or are touched in some other way by its effects.

A major contribution of this body of work is in its suggestion that the 'carceral' is something more than merely the spaces in which individuals are confined – rather, that the 'carceral' is a social and psychological construction of relevance both within and outside of carceral spaces. In order to highlight this development, the chapter first outlines the nature of the inside/outside binary as critiqued within carceral geography, then considers the discourses around the liminality of certain carceral spaces, the construction of carceral spaces 'outside' the institution, and finally explores the notion of the embodied reach of the carceral beyond formal spaces of incarceration, through a discussion of transcarceral spaces and inscribed bodies. In engaging with the corporeal experience of *release* from incarceration, it deals with a different embodied aspect of incarceration to that detailed earlier in Chapter 4. Finally, it draws together the scholarship on prison boundaries with wider geographical discourses of bordering and boundary studies, to suggest some future directions this area of inquiry could usefully take.

Challenging the Inside/Outside Binary

Considering first the prison wall and its effectiveness in delineating the prison from wider society, a starting point for much inquiry has been the work of Goffman on the 'total institution'. Goffman's classic definition describes '… a place of residence and work where a large number of like-situated individuals, cut off from the wider society for an appreciable length of time, together lead an enclosed, formally administered round of life' (Goffman 1961: 11).

In subsequent scholarship, the label of the 'total institution' has been applied to a diverse range of institutions and contexts, such as psychiatric units (Skorpen et al. 2008), the home (Noga 1991), homes for the elderly (e.g. Mali 2008), the mass media (Altheide 1991), the military and the police (Rosenbloom 2011), and sport (Cavalier 2011). While demonstrating considerable transferability and utility, the 'total institution' has been treated as something of a straw man in the context of the prison, and its applicability to this context has been widely critiqued, with Goffman's description of the total institution taken arguably too literally, and elements of his argument overlooked (Schliehe 2014). Farrington (1992: 6), for example, argues in relation to the US prison system that the 'total institution' thesis is 'in fact, fairly inaccurate as a portrayal of the structure and functioning of the ... correctional institution' in that the modern prison 'is not as completely or effectively "cut off from wider society" as Goffman's description might lead us to believe'. Goffman of course never intended to convey a sense of a hermetically sealed institution – he specified areas and circumstances where the institution and the outside world came into contact with one another. However, critiques of Goffman have value in that they highlight the nature of the prison boundary and open a space for debate over the kinds of permeability, porosity, transfer, exchange and other boundary-crossing processes which are observed to take place.

At the heart of Farrington's (1992) argument, and common to much subsequent work, is an understanding of prison institutions as having a relatively stable and on-going network of transactions, exchanges and relationships which connect and bind them to their immediate host communities and to society more generally (Farrington 1992: 7). Although at the time of writing, Farrington (1992) observed that relatively little research had explored these connections, such as the relationships between prisons and their host communities, the process of prison siting, and the relationships between criminal offenders and the society from which they have come, in the intervening years these topics have come more clearly into view. Farrington (1992: 7) essentially refined Goffman's notion of the prison as a 'total institution', suggesting instead that prisons resembled 'a "not-so-total" institution, enclosed within an identifiable-yet-permeable membrane of structures, mechanisms and policies, all of which maintain, at most, a selective and imperfect degree of separation between what exists inside of and what lies beyond prison walls'.

Baer and Ravneberg (2008) explicitly problematized the conceptualisation of a binary distinction between 'inside' and 'outside', positing the concept of heterotopia as a means of understanding the nature of the boundary, and viewing prisons as 'heterotopic spaces outside of and different from other spaces, but still inside the general social order' (Baer and Ravneberg 2008: 214). This understanding of the prison as still inside the general social order, they argued, renders problematic the separation of inside from outside. They built on Foucault's work, in which he characterised heterotopias as 'real places, actual places, places that are designed into the very institution of society' (1998: 178) but which can seem totally unrelated to one another despite existing side by side. In so doing, they relied particularly on Genocchio's (1995) observation that heterotopias are 'outside' of and fundamentally different to all other spaces, but also relate to and exist within general social space that distinguishes their meaning as difference. Baer and Ravneberg (2008) argued that the concept of heterotopia allows for a fuller understanding of the spatial complexities of the prison environment than the total institution thesis which distinguishes between inside and outside with very little room for blurring of this boundary. In their comparison of English and Norwegian prisons, they found what they described as 'incompatible juxtapositions' (ibid. 212), in which there were 'multiple, simultaneous distinctions and indistinctions' between the inside and outside of prisons, rather than a set of neat binaries. They described that they 'sensed a lack of delineation between inside and outside at the same time that there was sharp distinction within prison' (ibid. 213), and that prison seemed 'to be a compressed mélange of seemingly incompatible juxtapositions' (ibid.) with 'tension and fusion between inside and outside' (ibid. 214). These 'incompatible juxtapositions' derived from Baer and Ravneberg's own personal impressions of entering and leaving English and Norwegian prisons, or, to use Farrington's (1992) term, their individual 'interpenetration' of the penitentiary wall from a position on the 'outside'.

While Farrington (1992) identified 'points of interpenetration' through which the prison and wider society intrude into and intersect with one another, his theorisation still rested on an implicit understanding of the 'inside' of the prison and the world 'outside' as two entities. Whereas Baer and Ravneberg (2008) identified tension and fusion between inside and outside, they too saw the space of the prison and the spaces outside as existing in a binary relationship, even whilst they argued for the indistinction of the binary itself. Recognising the limits of their methodological approach, though, in their critique Baer and Ravneberg (2008: 214) stressed that further understandings of the nature of the prison, whether as a total institution or otherwise, should aim to extend beyond the perspective of the external observer, to incorporate the experience of the boundary by prisoners, whose perceptions of the 'inside/outside distinctions and indistinctions [may] take on different complexities and subtleties'. More recent work in this field, drawing on such perceptions, has suggested the existence of a third, 'liminal' space, overlapping and coterminous with the 'inside', but sharing characteristics with the outside.

Liminal Carceral Spaces

Theorisation of spaces which straddle, or blur, the prison boundary as 'liminal' draws on the work of Van Gennep (1909, translated into English in 1960), who first used this term to describe the transition from adolescence to adulthood in traditional societies, wherein the margin or the liminal is a space in which social rules are suspended because the subject no longer belongs to their old world, or to their new one – instead dwelling temporarily in 'nowhere land'.

Building on Van Gennep's work, Turner (1967: 81) developed the concept of liminality in more 'complex' societies, describing it as 'necessarily ambiguous, since this condition ... elude[s] or slip[s] through the network of classifications that normally locates states and positions in cultural space. Liminal entities are neither here nor there, they are betwixt and between the positions assigned and arrayed by law, custom, convention and ceremonial'. The word liminal, from the Latin *limen* which itself means boundary or threshold, has since been invoked in a variety of social and cultural contexts, and has been used by human geographers in combination with conceptualisations of space to describe specific spaces of betweenness, where a metaphorical crossing of some spatial and/or temporal threshold takes place.

For example, geographers have exploring hotels as liminal sites of transition and transgression (Pritchard and Morgan 2006), council tenants' fora have been viewed as liminal spaces between lifeworld and system (Jackson 1999), the liminal act of breastfeeding has been considered to demarcate specific spaces (Mahon-Daly and Andrews 2002), there are liminal notions of co-existence in Australia (Howitt 2001), the street has been researched as a liminal space for prostitutes in Brazil (de Meis 2002), and cyberspace has been viewed as a performative virtual liminal space for new and expectant mothers (Madge and O'Connor 2005). These authors have highlighted and problematized the 'betweenness' of the liminal, deploying the conceptualisation of the liminal as 'transformative' to frame the transitive experiences of individuals experiencing, shaping or creating these spaces. As Shields (2003: 12–13) has argued, many have highlighted the transformation of social status which is facilitated or permitted by liminal spaces, where 'initiates' are 'betwixt and between' life stages and where liminal spaces are bound up with ideas of becoming.

In Madge and O'Connor's work (2005), cyberspace allows expectant mothers to 'try out' different versions of motherhood, as they await the arrival of their babies. This notion of a performative process of experimentation with different maternal personae draws on Van Gennep (1960) and Turner (1967) in seeing liminality as a rite of passage between one world and another, and in particular Van Gennep's (1960) description of the three stages of passage; separation from a previous life, or the 'pre-liminal'; transition, the 'liminal', and reintegration in a 'new' life, the 'post-liminal'. Turner's (1967) work focussed primarily on the liminal stage, in which he described individuals entering an unstructured egalitarian world which he termed 'communitas', where comradeship transcends rank, age, kinship and so on, and displays an intense community spirit, in which social groups form strong bonds free from any structures which would usually constrain them. In the post-liminal (Van Gennep 1960) individuals leave communitas and reintegrate into their 'new' life, adopting a new social status and re-entering society in accordance with this new status.

The prison visiting room has long been recognised by criminologists and prison sociologists as a liminal space. For example, in her work on prisons in California, Comfort (2003: 80) described the visiting suite as a 'border region of the prison where outsiders first enter the institution and come under its gaze', and theorised this space as one in which visitors became subject to 'secondary prisonization' as a collateral effect of incarceration, 'a liminal space, at the boundary between 'outside' and 'inside', where visitors convert from legally free people into imprisoned bodies for the duration of their stay in the facility' (ibid. 2003: 86). Codd (2007: 257) has similarly described visiting as a 'liminal space' in which prisoner families 'are not entirely prisoners; however, they are within the prison establishment and thus defined as not entirely free either'. The spaces where prisoners and visitors meet are characterised by a 'liminal indistinction between inside and outside in which both the physical space and the experience of it are reflexively interrelated' (Moran 2013: 182). Different types of visiting spaces bring different elements of liminality to the fore; as Comfort notes, women participating in 'family visits' in bungalows within a patrolled compound rarely described feelings of confinement or captivity, whereas visitors entering cafeteria-style rooms where they sat at tables opposite prisoners, were made to feel 'as if they're incarcerated too' (Comfort 2008: 63). However, if the sense of liminal 'betweenness' experienced in visiting rooms by both prisoners and visitors is readily comprehensible, the notion of transformation which is also a characteristic of liminal spaces is less obvious.

The notion of a linear progression of transformation through distinct stages, as described by Turner (1967) has been contested, for example by scholars of disability who have argued that the permanently disabled may experience a state of permanent liminality where societies obstruct their social reincorporation (Phillips 1990, Willett and Deegan 2001). In relation to carceral spaces, for Russian women prisoners, the liminal space of the prison visiting room has been described as experienced not just once, as one stage in a linear transformation, but repeatedly, with the liminal coming to constitute a temporary, transient transformation followed not by a post-liminal reintegration into a different social status, but by a return to the state experienced before pre-liminal detachment (Moran 2011). Comfort (2008: 27) has similarly described visitors 'hovering between their outside lives and the inner world of the institution … [and] repeating the process each time they move through the intermediary space'. These observations point to the visiting room as a temporarily transitive space, entered and left by both visitors and prisoners who return to their pre-liminal context, rather than achieving some new post-liminal status. However, the repeated experience of prison visiting is also observed to have a subtle, cumulatively transformative effect. For Comfort, visitors in California are 'changed' by their experiences, in that 'recurrent exposure to this ordeal will itself become a transformative course' (ibid. 2008: 28). Prisoners themselves also change; women visitors may coax their [incarcerated] partners into 'being a "docile body" … who will "do his time" calmly and without incident and thereby minimise his risk of falling victim to violence or incurring disciplinary penalties that extend his time behind bars' (Comfort 2008: 186–7). For Russian prisoners, although the visiting suite is no immediate conduit to another life status, the cumulative effect of visitation is to remind prisoners of what life on the outside is like, and to motivate them to complete their sentences successfully in order to be able to return to it (Moran 2011). The space of the prison visiting room operates as a stage for partial and repetitive threshold-crossing, where transformation is both temporary and fleeting; but also subtle, cumulative and sometimes counter-intuitive.

Carceral Spaces 'Outside'

A third theme of work explores the idea of carceral spaces which extend beyond the prison itself, and also the implications of the embodiment of imprisonment for the 'reach' of incarceration beyond the point of release.

An example of research which theorises carceral spaces outside the prison is Allspach's (2010) study, in which by exploring the experiences of women released from federal prisons in Canada, she demonstrated that liberal 'welfarist' ideals embedded in neo-liberal reforms facilitate a network of social control, forming carceral spaces beyond prison walls which perpetuate and exacerbate marginality after release from prison. Focussing specifically on released women's own experiences of crossing the prison boundary from the supposed 'inside' to 'outside', she found that these women spoke of their post-release experiences essentially as another form of incarceration. Their 'outside' took the form of halfway-houses in deprived neighbourhoods, surveilled by closed-circuit television, in which their movements were micro-managed and monitored by parole officers and halfway-house staff. This 'outside', therefore, very closely resembled the 'inside' from which they had recently emerged as 'free' citizens. In fact their 'outside' experiences were so closely reminiscent of the surveilled institution of the prison, in that both their agency and independence were reduced, and their social contacts restricted, that Allspach describe them as 'socio-economic spatial re-confinements' (Allspach 2010: 720). This work recalls the argument made by Carlton and Segrave (2011: 552), that understanding women's experiences post-release requires analysis to extend 'beyond the walls of institutions' to adequately incorporate an appreciation of imprisonment as something other than a discrete traumatic episode in women's lives.

In a similar vein, Gill (2013: 26) explored the use of Electronic Monitoring as a natural extension of or alternative to mainstream carceral environments, suggesting that confinement can be independent of physical restriction, and drawing on Carnochan's (1998) observation that forms of punishment that are not explicitly prison-based can be just as constraining, in a different sense, as traditional incarceration.

Although for Allspach (2010), transcarceral spaces acted *after* release to recall the circumstances of incarceration, for women prisoners in Russia the notion of transcarcerality has been argued to apply equally well to spaces experienced *before* incarceration, in which rather than extending the reach of the 'carceral' post-release, prisoners are inculcated into carceral regimes *before* incarceration

(Moran et al. 2013b). They argued that the conditions of *etap* serve to acclimatize prisoners, both through the physical conditions of transportation, and the overt control exerted by the penal regime. Although the Russian example is arguably extreme, these spaces of transportation are common to all carceral regimes, as convicted prisoners travel from court hearings, or from remand prison, to the institution in which they will serve their custodial sentence, and indeed they take form in the multiple movements that some prisoners experience during their sentences, as they are transferred from one facility to another (Mitchelson 2013). They exist beyond the assumed 'boundaries' of the prison, as mobile spaces which operate to extend the spatial reach of the carceral institution in ways very similar to those described in Canada by Allspach (2010).

In the Russian prison system, where prison transportation covers very large distances and for some individuals can take several weeks (Moran et al. 2013b), the testimony of women prisoners has enabled *etap* (also discussed in the previous chapter in relation to mobility) to be theorised as a mobile space in which convicted prisoners experience a sense of transformation from free individuals to prisoners, within their particular penal context.

Kept in cellular confinement in overcrowded train wagons without windows to the outside, prisoners are under direct surveillance by armed guards, and many are shocked and frightened by the experience. Prisoners described the adaptations made to the train carriages in which prisoners were transported, including the addition of bars and locks to keep prisoners spatially contained, and the presence of guard dogs and armed convoy staff. Prisoners were witness to transgression of certain rules about the conduct of prison transport and the segregation of categories of prisoner, and their sense of powerlessness was exacerbated by the fear of violence from convoy guards. One respondent recalled:

> They brought us to the station, it was cold, winter, and we were sitting in the freezing cold for an hour and a half waiting for the train. Then the train came, the first and the second [carriages] went to the men, and then next the women. There were 4-person coupés [compartments], but we had 10 people in them, with all their stuff in the compartment. We had kids with us too, and they were going further than us. Despite the fact that they are kids, they travelled with us. They were between 14 and 18. Although they're meant to go separately they went with us. The HIV-infected should go separately, people with TB must go separately. They went together. And who can we complain to? (Moran et al. 2013b: 116)

Etap, with its spatial context of the patrolled and secure train carriage, and the overt and punitive control exerted by convoy personnel over transported prisoners, has previously been argued to serve 'to preface, or anticipate, the experience of incarceration for Russian prisoners, extending the penal regime of the prison into these mobile "transcarceral" spaces which precede incarceration "proper"' (Moran et al. 2013b: 117). *Etap* is transformative in that it prepares prisoners for what is to come. As a stage leading to incarceration, *etap* could also be seen as a circumstance of betweenness – although convicted, those being transported, whilst not yet incarcerated, are neither free citizens nor inmates. *Etap* also resonates with Turner's (1967) description of individuals entering an unstructured space in which they form social bonds and live by new and different rules (Moran et al. 2013b). Although the spaces of *etap* are formally organised and managed by the prison system, informal interactions take place both between transported prisoners and between prisoners and guards, and this shared experience in the context of the isolation and destabilisation characteristic of *etap*, allows prisoners exiting this liminal space to reintegrate into their 'new' life (Van Gennep 1960) as prisoners.

The liminal transcarceral space of transportation tends to be characterised by the isolation of individuals, to separate them from their pre-liminal context and to prepare them for post-liminal integration into some new situation. When a Russian prisoner receives a custodial sentence, they commonly remain unaware of where that sentence will be served either until *etap* begins, or occasionally until arrival at the prison itself (Moran et al. 2013b). They seem to 'disappear' within *etap*, hidden from contact with friends and family for the duration of a journey which can take weeks or months. As one prisoner recalled 'No one told me where I was going. My father and brother didn't even know. My uncle found out through the police. I'd been [in the destination institution] for a month before my family knew where I had gone' (Moran et al. 2013b: 118).

If, as discussed earlier, the facility for friends and family to visits loved ones during incarceration presents the opportunity for a liminal space to be created inside the prison, then the prison furlough, or home visit, creates another carceral space of indistinction on the outside. By allowing prisoners to leave the physical confines of the penitentiary, the space of furlough could be conceptualised as a simultaneous distinction and indistinction of inside and outside; a space outside the prison which offers freedom from carceral control, but to which access is strictly controlled by the penal authorities.

Research in Finland (Moran and Keinänen 2012), where furlough is commonplace and unremarkable, has shown that the selection of prisoners for furlough by the penal authorities rewards prisoners' 'docility', in terms of their compliance with the carceral regime during incarceration. In deciding which prisoners are be allowed to traverse the boundary of the prison 'proper', they tend to reward good behaviour in prison, and to view older prisoners, first-time inmates, women, Finnish nationals, and married prisoners as at lower risk of breaching the conditions of furlough. In this selection process, prisons arguably blur the distinction between the inside and the outside of the prison by selecting candidates who they perceive to be more likely to sustain the good behaviour demonstrated inside the prison, on the outside. In so doing, they hold the outside and inside in tension and fusion with one another, granting access to the outside on the basis of good behaviour, and encouraging the spatial translocation of the 'docility' demonstrated inside the prison. Or in other words, that they attempt to create, through the case-by-case selection of prisoners for furlough, a transient space that is simultaneously 'outside' of the prison, but which when properly observed, exists both 'within' the prison (since the prison sentence continues to be served), and within the general social order of the 'outside' (after Genocchio 1995, in Baer and Ravneberg 2008). The extremely low rate of breach of furlough conditions in Finland (less than 5 per cent across the entire prison population) suggests that this selection procedure is relatively successful.

The Embodied Reach of the Carceral[1]

Early theorisations of situational adjustment to imprisonment (Becker 1964, Wheeler 1961) tended towards an understanding that formerly incarcerated people 'may be insulated from lasting socialisation effects' (Wheeler 1961: 711). However, a notion of the effects of incarceration reaching beyond the prison wall is now well established within criminological literatures (Garland 1990, Mathiesen 2000, Sykes 1958, Toch 1977), and has been explored, for example, in relation to prisoner mental health (e.g. Tye and Mullen 2006, Douglas and Plugge 2008), prisoners' families and children (e.g. Hannon 2007, McAlister et al. 2009), and the challenges of resisting reoffending (e.g. Maruna 2001, Sampson and Laub 1993). Where this discussion departs, and where carceral geography can make a distinctive contribution, is in its explicit consideration of the *embodied* nature of this extended reach; the way it is entwined with the incarcerated and post-incarcerated body in complex and nuanced ways.

Extending the discussion above, and the discussion of the embodied experience of incarceration presented in Chapter 4, transcarceral spaces arguably exist alongside and perhaps also in combination with, an embodied sense of the 'carceral' which is similarly mobile beyond the prison wall through the corporeality of released prisoners. Recalling the earlier discussion of the embodied nature of the experience of incarceration, specifically the corporeal inscription of imprisonment and the embodied strategies adopted by prisoners, this section of the chapter considers the ways in which this embodiment of incarceration reaches beyond the physical space of the prison.

Drawing on feminist scholarship to conceptualise the experience of incarceration as inherently embodied, embodied inscriptions and strategies of incarceration can be read as corporeal markers of imprisonment, blurring the boundary between 'outside' and 'inside' the prison and extending carceral control. The insights derived from an embodied perspective on experiences 'inside' prison offer much

---

1   Drawing on Moran, D. 2012. Prisoner Reintegration and the Stigma of Prison Time Inscribed on the Body. *Punishment & Society* 14: 564–83 and Moran, D. 2014. Leaving behind the 'Total Institution'? Teeth, TransCarceral Spaces and (Re)Inscription of the Formerly Incarcerated Body. *Gender, Place & Culture* 21(1): 35–51. Research data were generated through the ESRC project *Women in the Russian Prison System* conducted with Judith Pallot and Laura Piacentini.

for understandings of experiences 'outside', and indeed for the question of whether the experience of individuals released from incarceration can be characterised in terms of this binary distinction. Specifically, considering the continued 'control' of released prisoners, a feminist perspective of embodiment opens a space for discussion of how that control is felt personally, its manifestations, and the ways in which those affected internalise or contest the extended reach of the carceral into their lives on the 'outside'.

Released prisoners face the challenges of 'reintegration', into society after release, trying to reconstruct their family lives, to reacquaint themselves with parents, partners and children, and critically, trying to find employment. They must try to cope with life after release, getting used to 'normal' life and adapting to the changes which have taken place in their absence. Reintegration has been intensively studied, particularly by criminologists and prison sociologists, and while a detailed consideration of this research is beyond the scope of this chapter, a brief survey of studies as they pertain to embodied experience is presented here. Researchers have in general sought to identify the critical factors contributing to successful reintegration (broadly defined as former prisoners functioning as members of mainstream society rather than reoffending and being reincarcerated). A survey of this scholarship shows that the key, interconnected, factors include; finding employment on release (and thereby overcoming the stigma attached to ex-offenders entering the labour market); maintaining good health (to contribute to a good quality of life and to enable employment); maintaining connections to family and community (for both emotional and financial support and also accommodation); obtaining secure and affordable housing (again related to employment or income); obtaining and retaining appropriate documentation, and avoiding substance abuse (Travis 2005). Broader contextual issues include the maintenance of community supervision, partnership working between local stakeholders, and public safety (e.g. Petersilia 2001, Travis et al. 2001, Travis 2005).

Despite the attention paid to prisoner reintegration, as a critical time during which, it is perceived, the right kind of assistance can prevent recidivism and improve outcomes for the previously imprisoned, very little attention has been paid to the physical manifestations of imprisonment, in terms of the embodied experiences of imprisonment and the ways in which these impact on life after release; in particular, the specific ways in which they might impact gendered bodies leaving carceral space. Studies of barriers to participation in the labour market, for example, while they highlight the problem of stigmatisation for former prisoners (the 'prison effect'), tend to focus on the formal disclosure of a criminal past (e.g. Weiman 2007), rather than on the subjective, personal judgement of individuals based on their appearance. Former prisoners' corporeal bodies have thus far come under scrutiny only as vessels of illness or disease, when authors consider the considerable challenges faced by prisoners released in poor health and requiring expensive healthcare services for which they cannot afford to pay, or where illness renders them unable to work (e.g. Hammett et al. 2001, Mallik-

Kane 2005, Davis and Pacchiana 2004), rather than in a way which enables bodily subjectivities to be recognised and considered.

In her work on the reintegration of women prisoners, Zaitzow (2011: 209) pointed out that 'what happens inside jails and prisons does not stay inside jails and prisons'. She provided a grounded and nuanced overview of the particular challenges facing women on release from prison, including the stigma attached to incarceration, which translates into an embodied notion, with women saying that 'they believe they have a tattoo on their forehead that proclaims them as "ex-con"' (ibid. 242). However, her work posed further, critical questions about exactly *how* this embodied experience of imprisonment works to stigmatise those released from prison. Using empirical evidence gathered during research with women released from prison in Russia, and drawing on research with men released from prison in the United States, this chapter goes some way towards providing an answer to these questions.

Transcarceral Spaces and Inscribed Bodies

In communities where incarceration does not form part of the majority experience, a personal history which includes incarceration is a source of stigma, with former prisoners facing significant obstacles to assuming mainstream social roles. Incarceration is generally perceived to carry a stigma that marks the previously incarcerated as dishonest or unreliable, often seen in labour market studies where employers express a strong preference against hiring former prisoners (Lopoo and Western 2005). Although the stigma attached to incarceration is well understood in terms of its operation and its effect, i.e. the stereotypical views which are held about former prisoners, and the separation, status loss and discrimination which they suffer as a result, much less is known about how those affixing stigma to previously imprisoned individuals know to do so. This status is not tattooed on foreheads, but according to the experiences of female former prisoners interviewed in Russia, something about their bodily subjectivity marks them out.

Formerly incarcerated women frequently spoke about their sense that their experience of incarceration was plainly apparent to others – that people would be able to 'tell', to discern from their appearance that they had been to prison. As one woman explained:

> I was out [of prison] for two months before I would come out of the house. I was afraid that "zek" [a Soviet era word for prisoner] was written all over my face. I was afraid of people.

Another former prisoner initially kept her history a secret from work colleagues, for fear of their reaction, disclosing it to them on the anniversary of her release. Her recollections of this event revealed their preconceptions.

> When I came to work here we [the former prisoner and her work colleagues]
> drank tea together and got to know each other … When I got my first paycheque,
> I didn't say anything to anyone, just made tea and said – "Girls, let's have a
> cup of tea". When everyone was sitting down and drinking tea, I said "You can
> congratulate me", because it was October 5th. I said "Girls, I want to share my
> joy; it's a year since I was released [from prison]". It was such a shock for them,
> all sitting there! … They were all like "oh-oh-oh", laughing. And they were
> like, "Tatiana, you're lying!" – "You don't smoke, you have no tattoos and you
> don't swear!"

As this extract shows, although her work colleagues were supportive when
she disclosed her history to them, their reactions revealed both their negative
preconceptions about former prisoners, and the fact that they perceived prison
to 'mark' women in a certain way – through their physical appearance and
personal habits. In their minds, smoking, tattoos and swearing were human
differences which could be clearly observed, and which indicated that a person
could previously have been incarcerated. Another formerly incarcerated woman
expressed this much more starkly.

> They drink and smoke, they have tattoos, and they use slang. And their teeth are
> rotten. That's how people think women who've been in prison look.

For these women, prison time is clearly inscribed on their bodies through loss of
teeth. They feel conspicuous and different as a result, and perceive themselves to
be easily recognisable as former prisoners. Missing teeth are, to their minds, the
human differences which are noticed by others, allowing them to link dominant
cultural beliefs about the previously incarcerated to these labelled differences
through negative stereotyping.

It is, of course, possible that a different process is in train here, that persons
affixing stigma to formerly incarcerated individuals on the basis of their missing
teeth do so on the basis not of dominant cultural beliefs about the formerly
incarcerated per se, but instead in relation to other stereotypes about people
with poor dental health (Fisher et al. 2005). Whatever the precise mechanism
through which this happens, what is clear is that women experience shame and
embarrassment about their teeth. They feel that their bodies do not 'fit' the spaces
in which they find themselves (McDowell 1999), both in a general context in
which Russian women are increasingly subject to aspirational physical ideals; a
'Barbie doll image of beauty' (Kay 2000: 90), and specifically in a job market
in Russia which privileges physical appearance (Moran et al. 2009), particularly
in the retail jobs to which many of these women aspire; a context which induces
shame and guilt about the less-than-ideal body. As Yakubovich (2006) has noted,
in the circumstances of post-socialist transformation, Russian employers have
formally and routinely indicated ascriptive characteristics such as age, gender,
health and physical appearance requirements in job advertisements, in addition

to the informal discrimination which pervades most labour markets (Hamermesh and Biddle 1993). Their comments, discussed below, describe compellingly the simultaneously very personal and overtly public nature of missing teeth as prison time inscribed on the body in a way that was very embarrassing, and very difficult to conceal.

This particular corporeal inscription of incarceration, in terms of the loss of teeth, is unlikely to be either intentional on the part of the Russian prison system, which attempts to provide adequate medical and dental care to all inmates, or directed uniquely towards women. A comparison of the experiences of male and female ex-prisoners is beyond the scope of this discussion, but it is likely that the apparent institutional neglect of dental care is replicated in men's prisons. While the loss of teeth is therefore unlikely to be inherently gendered, the women's testimony suggests that women's embodied experience of this inscription is distinctive, in terms of the ways in which the 'lack of fit', as theorised within feminist geography (McDowell 1999: 61), between their formerly incarcerated bodies and the circumstances in which they find themselves after release are played out; in particular the adverse effects of this inscription for their entrance into and participation in the labour market. In line with the observations made by Williams (2007) for the US, these Russian women found that deterioration in their dentition caused problems on release from prison, particularly when trying to find work, and they articulated these concerns by contrasting their own appearance with that of younger and more physically attractive or 'appropriate' competitors in the job market.

Q. And did they not do your teeth in prison?

A. No, that's the worst thing. There, all the girls, all of them, they come out toothless – Not one will give you a job [on the outside]. Not when you open your mouth.

I had to get my teeth done. I got my front teeth done before anything else. I couldn't even open my mouth or, well, who would give me a job? It's easier for them to take a young girl who looks good in the same shop, than me with no teeth.

In her work on the gendered embodiment of incarceration for formerly imprisoned men in the United States, Caputo-Levine (2013) found that the embodied strategies of carceral habitus adopted in response to the threat of interpersonal violence in prison, remained with the individual to a greater or lesser extent after release. In her example, the conduct of men recently released from prison was characterised by a tendency to reproduce on the outside the gendered behaviours necessary to maintain personal safety in the hyper-masculine prison environment.

She noted that former inmates now attending career development classes on the outside, to assist with finding employment and negotiating job interviews, still employed elements of the carceral habitus. Those men recently released were identifiable by their well-developed upper body, their particular manner of holding the body, and of holding the body in space, which conveyed an ability to defend one's self, if necessary. At times of stress, such as during mock job interviews, some former prisoners adopted the blank 'yard face' which functioned in prison both as an expression of aggression, and to enable the individual to withdraw and assess a situation whilst concealing their intentions. Entering a room for skills training, former prisoners would survey the room, and in line with a hyper-sensitivity to physical vulnerability, would choose seats with their backs to the wall. Those most recently released experienced great difficulty in dropping the 'yard face' and smiling at others – a recollection of the fact that 'in the milieu of the hypermasculine prison a smile is seen as declaring yourself to be open to attack' (Caputo-Levine 2013: 176). An acute sensitivity to perceived disrespect, leading to snap reactions and sometimes physically and verbally threatening escalations, led to problems both in everyday life (such as in accidental bumping incidents on crowded subways) and in career development classes, as did an inability or reluctance on the part of parolees to engage in small talk with people they had just met, or indeed with classmates who they may have known (or feasibly may expect to meet again) in prison. All of these aspects of carceral habitus acted to inhibit the reintegration of former prisoners, and their ability to find and keep employment, contributing to isolation in the workplace and to negative interactions with employers and authority figures.

Overcoming Stigma: Reinscribing the Formerly Incarcerated Body

Women interviewed in Russia were willing to go to considerable trouble and expense to repair or replace their teeth, and they devoted their scarce resources as a high priority on release.

> Here [outside of prison], you have to get your teeth done, above all. Now you can have them put in [implanted], my mother gave me 56 thousand rubles [approximately 2000 US dollars] to have teeth put in.
>
> Q. That's expensive.
>
> A. Yes. [But] I had no teeth, I was ashamed to open my mouth.

Russian prisoners face a variety of obstacles to successful reintegration, including the renewal of essential personal documents like the internal passport, residence permits, and so on, with implications for their legal status and entitlement to various aspects of citizenship. What was clear, though, was that the stigma attached to their teeth made them deal with this issue more urgently than almost anything else, in an attempt to rectify the 'lack of fit' (McDowell 1999: 61) between their own bodies and the idealised representations expected in the post-carceral spaces of employment.

> I'm not in a hurry to get my documents, and I haven't visited the doctor's. The only thing is – dentistry – that's the first thing I'll do when I'm more or less back on my feet.

Having identified the problems they face in the labour market with poor dentition, it follows that the inability of certain women to pay for remedial dental treatment reinforces the gender-class-producing function of carceral inscription through hindering their pursuit of employment, and thereby reducing their likelihood of reintegration. Restricted in their employment opportunities, women with poor dentition are perhaps more likely to find themselves in lower-skilled, non-public-facing employment such as cleaning or factory work, where their prospects for career progression are minimal.

Similarly, working in the United States with released men whose embodied carceral habitus shaped their post-release identities, Caputo-Levine (2013: 178–9) observed that some former prisoners recognised their own semi-conscious adoption of the carceral habitus, and tried to modify it. She discussed the example of Marley, who by watching the ways that others responded to him, started to make changes to his appearance and the way in which he carried and conducted himself. Realising that others were afraid of him, he changed his mannerisms and the way that he walked, practising being non-threatening. He even went so far as to purchase spectacles with non-prescription lenses, which he felt put people at their ease when speaking to him.

By demonstrating a desire either to repair the damage done to their appearance in prison, or to throw off the carceral habitus they had adopted in prison, in both the United States and Russia, these men and women continued the unfinished, gendered projects of their bodies. In this way, they show that as Wahidin (2002: 192) has argued, the material body is not a passive receptacle for inscription, but rather is interwoven with and constitutive of systems of meaning, signification and representation. They treated their bodies as projects, always in the process of becoming through the experiences of embodiment.

Conclusion and Directions

The assumed binary between 'inside' and 'outside' of prisons has already been heavily critiqued, and in drawing out the threads of the critical discourses, this chapter has drawn attention both to the indistinction of the prison boundary itself, and to some of the spaces which although nominally lying 'within' prison, also act as liminal sites of boundary crossing; and also to some of the diverse sites and contexts 'outside' of prison in which aspects of incarceration are reproduced and replicated. The notion of the 'carceral' as a social construction existing both within and separate from physical spaces of incarceration can be developed and extended further, in ways which will assist in the burgeoning theorisation of the nature of the carceral boundary.

What is clear from existing studies of prison boundary crossing is that the prison wall itself is porous, permeable, interpenetrated, and that it does not spatially demarcate the limits of the prison in terms of the experience of incarceration. Exactly how the contested prison boundary is to be theorised and understood, though, is still under debate. Although liminal carceral spaces are argued to exist within the prison, transcarceral spaces outside of it, and an embodied notion of stigma is observed amongst former prisoners, the fact remains that the prison is still demarcated from wider society in ways which render its walls selectively impenetrable. Although carceral geography has made significant advances in understanding the contested nature of the prison boundary, there is still work to be done.

The critique of the inside/outside binary of the prison boundary in carceral geography recalls the 'seductive charm of the border' which has preoccupied critical border studies within human geography (Parker et al. 2009: 584). In this field, the inside/outside binary is recognised as a trope within Western thought which tends towards such binary oppositions, and which functions to obscure indeterminacy and indistinction. In border studies, the binary of inside/outside has been closely associated with the idea of nation state and territory, and accordingly, there is a movement towards attempting to disassociate the border from territory, in order to enable alternative topologies to be explored, which make it possible to conceive of the border 'within a more relational understanding of difference' (ibid.). Although carceral geography has already achieved a sense of this dislocation of the prison border from physical territory, in part through the consideration of embodied experiences, transferring ideas from critical border studies may be a means to further advance scholarship in this area. The intersection of carceral geography and critical border studies has already been observed around the confinement of migrant detainees, where the border of the nation state is no longer found only at territorially identifiable sites such as ports and airports, but is also transferred to the facilities which hold noncitizens awaiting decisions on their entry or deportation. However, this synergy need not be limited to migrant detainees; there are some useful parallels with the experience of 'mainstream' prisoners crossing prison borders in relation to legal status, disenfranchisement, and restriction of citizenship.

Specifically, carceral geography should 'borrow' from critical border studies in order to take a more experiential approach to the prison boundary, for example in terms of the theorisation of the phenomenological dimension of the border; and rather than thinking of it as a static line, conceive of it as a series of bordering *practices* which are embodied, multisensory, mobile, emotionally charged, and both spatially fixed and despatialized.

PART III
The Carceral and a Punitive State

Chapter 9
The Carceral and a Punitive State

The first two sections of this book have explored the nature and experience of carceral spaces, and the geographies of carceral systems. So far the spatiality of imprisonment has been explored at the level of the individual, focusing on the individual embodiment of incarceration, the experience of the passage of time; and in relation to (networks of) institutions, looking at the logics of prison siting, and the nature of the mobilities inherent in prison systems. Underpinning these concerns, however, at a higher level of spatial and conceptual abstraction, is the relationship between the socially constructed notion of the 'carceral', and the state, and specifically the symbiosis argued to exist between spaces and practices of incarceration (whether 'inside' or 'outside' prisons) and, in certain contexts, an increasingly punitive state. Opening the third and final section of the book, this brief chapter therefore explores this relationship between the carceral and the state, paying particular attention to the context of the 'new punitiveness' (Pratt et al. 2011).

This chapter unfolds as follows. Dealing first with the notion of the 'new punitiveness' and the prevalence of this understanding of current penal policy within relevant literatures, it then briefly reflects on the discourses of hyperincarceration and the carceral churn, and considers the ways in which the social construction of the 'carceral' reflexively shapes and is shaped by the changing state and its philosophies and policies of punishment. Moving next to consider the work of human geographers engaging with these discourses, it concludes by prefacing the two subsequent chapters exploring the design of prison buildings and the interpretation and experience of carceral cultural landscapes and 'post-prisons'. The intention here is to explore relationship between the 'carceral' and the state, as it is manifest in the built form of prisons, as a mechanism through which the goals of a criminal justice system are materially expressed.

The 'New Punitiveness', Hyperincarceration and the Carceral Churn

Touched upon in Chapter 2 in relation to the origins of and dialogues within carceral geography, the 'new punitiveness' is a term given to describe the current *milieu* of penal policy in the United States, in many other English-speaking countries, and to a limited extent in Europe. Although penal policy changes over time in all jurisdictions, it has been argued that the 'new punitiveness', with its increased criminal sanctions, longer sentences, sense of 'shaming' and humiliating punishment, and retreat from discourses of rehabilitation, reaches beyond the bounds of the normal policy 'ebb and flow' (Pratt et al. 2011: xii, Moore and Hannah-Moffat 2011: 85). The wider trend towards US-style mass incarceration has triggered interest in the relative punitiveness of different countries and jurisdictions, bringing into view the relationship between incarceration and the nature of the state itself, with scholars identifying distinctive cultural, historical, constitutional, institutional and political factors that either facilitate or hinder the development of punitive policies (Gottschalk 2009: 440).

The work of social theorists and geographers such as Wacquant (2010a, b and c), Gilmore (2007) and Peck and Theodore (2009), has called for greater attention to the causes of and solutions to hyperincarceration (Wacquant 2010b: 74), 'prisonfare' (Wacquant 2010c: 197), and the carceral 'churn' (Peck and Theodore 2009: 251). This notion of a carceral 'churn' is a useful starting point, since it focuses discourses around reoffending and reintegration into a consideration of the relationship between the 'carceral' and a punitive state, drawing together themes related to prison/community relations and inside/outside which have been explored in previous chapters.

The carceral 'churn' is the repeated release, reoffending, re-sentencing and re-imprisonment of former prisoners, argued to operate to deliver the high rates of reoffending and re-imprisonment amongst prisoner populations in a variety of contexts. This 'churn' operates through a complex web of interconnecting factors pertaining to crime rates, sentencing policy, policy context, stigma, and so on. It is argued to create 'career criminals' who are increasingly difficult to 'rehabilitate' and who treat imprisonment as an 'occupational hazard', to give rise to generations of children for whom contact with the prison system becomes the 'norm', and to contribute to the development of communities in which criminality becomes endemic. Although reoffending and reimprisonment are the processes through which the carceral churn operates, the fundamental underpinnings of this system are argued to rest on the relationship between class transformation, ethno-racial division and neoliberalism, and the twin concepts of 'workfare' and 'prisonfare'. Loïc Wacquant's work is at the heart of this understanding of the penal *milieu*. He posited that selective reductions in state spending and privatisation of state activities do not mean that neoliberalism should be interpreted as the 'retreat' of the state; rather, he argued that the state has deployed a new range of strategies of social control and coercion.

He advanced an argument in three aspects. First, decoupling the assumed linkages between crime and incarceration, he argued that the expansion of the penal system is a response to social insecurity, rather than a reaction to trends in criminal behaviour. The US held 21 people for every 10,000 crimes in 1975, but 125 by 2005, indicating an increasingly punitive response to crime which in turn demands that attention be paid to the extra-penological functions of prison; in other words, what *else* prison is useful for other than removing and punishing offenders. Wacquant (2011a) contended that prison is an instrument of control of disruptive, discontented, impoverished ethnic groups and lower classes, heightened by state responses to social and economic changes; 'Police, courts and prison have been deployed to contain the urban dislocations wrought by economic deregulation, and impose the discipline of insecure employment at the bottom of the polarising class structure' (Wacquant 2011a: 205). In the United States, he argued, the swift ascent of the penal state has been driven by advanced marginality which is prevalent, entrenched and concentrated; a unique ethno-racial division which marginalises African Americans and Latinos (see also Alexander 2010); extreme social inequality with class and ethnic segregation in the metropolis; the commodification of public goods; religiously inflected moral individualism and pernicious notions of the poor as 'undeserving'. The non-response of the US welfare state to the racial crisis of the 1960s and the economic turmoil of the 1970s, he argued, left a space for the penal wing of the state to expand in response to social insecurity, aiming punishment at a precarious and stigmatised population.

Secondly, Wacquant (2009) suggested that political restructuring at the foot of the social and urban order delivered a situation in which the downsizing of public aid and the upsizing of the prison can be viewed as two sides of the same coin, or in other words, that the same resentful and racialized view of the poor has informed a punitive 'turn' in both welfare and justice policy. In this perspective, the poor are seen as 'depraved', not deprived; in need of correction, not support. Welfare and criminal justice are two modalities of public policy towards a population whose demographics of public aid recipients and inmates are almost identical. Originating from the same unskilled, marginalised, working class populations in the same urban neighbourhoods, these populations are the primary targets of what he called the 'double disciplining' of workfare and prisonfare.

For Wacquant, prisonfare in the US is the expansion of police, courts, prisons and their extensions such as probation and parole to over 5 million incarcerated people, 2 million people on parole, and 30 million on criminal databases. Enabled by criminal profiling and background checking, heightened fear of crime driven by media sensationalism, and the extension of criminal control to other branches of the public sector, prisonfare sits at the intersection between social and criminal policy, restricting opportunities for those in contact with either system. For example, as he observed, on release from prison, drug convicts are barred from living with their family if the family is accommodated in public housing, and employment opportunities for released prisoners are severely restricted by the necessity of background criminal record checks and the regulation of the employment of ex-convicts.

Finally, he posited that not only does the state shape the carceral system, but 'the meshing of workfare and prisonfare contributes to the *making* of the neoliberal state' (2011: 210, my emphasis). Rather than the neoliberal state retreating, it operates liberally at the 'top', in terms of a *laissez faire* attitude towards corporations and elites, arguably at the level of the causes of inequality, but is aggressively interventionist and authoritarian at the bottom of the class and status spectrum, where inequality becomes destructive. Wacquant essentially conceived of the typical neoliberal state as inherently penal, developing 'the punitive containment of urban marginality' (Wacquant 2010c: 210). In such a state, the most socially and economically marginalised classes are controlled through a mixture of prisonfare and workfare, with prisonfare acting as a system of 'warehousing' permanently unemployed sections of society, effectively penalising poverty; and with workfare seeing welfare rights becoming conditional on job-seeking at low wage levels.

Wacquant's thesis was developed primarily in relation to the United States, a factor at the heart of some of its critiques. An over-reliance on the neoliberal state as exemplified in the US, arguably acts to underplay the historical, political and cultural differences between countries. It is further argued that the thesis lacks a gendered perspective to understand how social policies escalate penal outcomes for women, and that it tends towards a view of the 'precariat' (Standing 2011) as largely passive, a collective body of victims unconcerned with autonomous struggle and political disorder.

Whilst perhaps of limited relevance for less punitive contexts, Wacquant's work has considerable resonance for the UK. In the British welfare state, as exemplified through current coalition government welfare policy, income maintenance, housing, education, immigration, health and social services (rather than mass incarceration) are arguably the pre-eminent means deployed to control, regulate and remake welfare-dependent 'problem populations' (Martin and Wilcox 2012). British political and popular media discourse about the 'precariat' plays a key role in the state's production of social disadvantage: the values and habits of working class families are portrayed as shameful and dysfunctional for entertainment and titillation, and to invoke anger and indignation, whilst obscuring the underlying causes and contexts of the social problems portrayed. Policy making is hostile towards the socially marginalised, responsibilising them for their 'deviancy', and demanding accountability for errant behaviour (Hancock and Mooney 2012). Violence and breakdown often accompany both advanced marginality and state attempts to re-impose forms of order and authority, for example in post-2011 riots Britain (Squires 2012).

It is important to note that although Wacquant's work has been prominent in the debate over the emergence of the new punitivity, with Gilmore (2007) building on his arguments to posit that the expansion of the US carceral estate functions as a spatial fix for the structural instabilities of surplus land, labour and capital, this is not the only explanatory framework advanced. As Kaplan-Lyman (2012) noted, whereas Wacquant and Gilmore see hyperincarceration as a *structural* element of the neoliberal state, others have tended towards a view of neoliberal *ideology* as the justification for an expanded criminal justice apparatus. For example, Harcourt (2008) has argued that a discourse of neoliberal penality has facilitated the rise in incarceration by allowing state intervention to focus on enforcement whilst being derided as inefficient in other spheres of activity. As he observed; 'the government is pushed out of the economic sphere, relegated to the boundary, and given free rein there – *and there alone* – to expand, intervene, and punish, often severely' (Harcourt 2008: 2). This view accords with Simon's (2007) contention that crime has become the central justifying logic for government intervention, in an era of neoliberal state retreat from issues of economic inequality and social welfare, leading to increasingly punitive policies on crime and punishment. Others have argued that neoliberalism creates a cultural logic that privileges individual responsibility, enabling a shift in welfare policy that make welfare recipients responsible for their own poverty, and underwrites punitive policies for criminal behaviour through the rejection of understandings of crime as connected to contextual circumstances (e.g. Katz 1996).

Human Geographies of the Prison/Ghetto Symbiosis and Urban Marginality

The relationship between spaces and practices of imprisonment and a punitive state has already received attention within human geography. Peck (2003) and Peck and Theodore (2009) have discussed the relationship between prisons and the metropolis in the context of hyperincarceration, in the aftermath of what Wacquant (2011: 3) described as 'a brutal swing from the social to the penal management of poverty' particularly in the United States, with a 'punitive revamping' of public policy tackling urban marginality through punitive containment, and establishing a 'single carceral continuum' between the ghetto and the prison (Wacquant 2000: 384). More recently, scholars have drawn attention to specific facets of the relationship between the 'carceral' and the state, with Mitchelson (2013) analysing the political controversy over the counting of prisoners in the US Census – where prisoners are 'counted' as residents of the prison where they are held rather than as residents of the places from which they have come, with the result that prison populations are used for political purposes (such as the allocation of federal funds) from which disenfranchised prisoners are generally excluded. Similarly, Nowakowski (2013) has drawn attention to the use of US prisoner labour, unprotected by workers' rights or health and safety inspections, to conduct hazardous waste-processing activity in ways which both subsidise the costs of imprisonment, and undercut private recycling companies.

Prisonfare also has the potential for considerable traction with urban geography, in relation to marginality and ethnicity. In their work on 'carceral Chicago', Peck (2003) and Peck and Theodore (2009) emphasised the flows into and out of prisons, prison expansion, ethnicity, and the cost of incarceration, finding that certain neighbourhoods experienced at first hand high numbers of released prisoners returning to impoverished communities with few opportunities, high levels of reoffending and the 'normalisation' of criminality, and an increasingly symbiotic relationship with the prison system; a revolving door between incarceration and impoverished communities. For Cook County, the jurisdiction which includes Chicago itself, they observed that there were very few employment opportunities, making crime one of the few occupations which *did* pay. Accordingly, the prison system itself became a labour market institution, somewhere to 'put' the unemployed by concealing or relocating unemployment, and reducing the labour supply, thereby reducing unemployment and reducing welfare spending. Whilst there are dubious short term benefits from this system, it has a pernicious long term effect on the released, who find it very difficult to find legitimate work, being caught in a 'carceral churn' between unemployment and incarceration.

Similarly, Shabazz' (2009) work on the Chicago Southside argued that the deployment of carceral spatial formations and techniques into the everyday lives of poor Black communities prepared and predestined many Black men for prison life. He argued that the combination of a bleak, monotonous, and poorly maintained urban environment, coupled with an overriding sense that public housing should not be comfortable, (in line with ideas of the 'undeserving' poor), against a backdrop of fundamental ethnic segregation, poverty, social and gang problems, meant that the security infrastructure installed in public housing, such as barred windows and turnstiles, vividly recalled carceral spaces, and thus acclimatised young men to imprisonment.

Directions

The preceding chapters of this book have explored individual experience of carceral spaces, the processes and logics of prison siting, and the idea that the prison 'wall' is porous in a variety of significant ways. Underpinning all of this work is a notion of the relationship between the carceral and the state, in terms of incarceration itself serving the perceived needs of the state in providing an 'appropriate' response to offending behaviour, translated into prisons which are located according to prevailing spatial logics specific to their socio-economic and political context, and which are experienced spatially by those accommodated in them. Although Philo (2012: 4) described carceral geographies as a sub-strand of 'geographical security studies', drawing attention to consideration of 'the spaces set aside for 'securing' – detaining, locking up/away – problematic populations of one kind or another', arguably there is more at play than this.

The aim here is to renew a focus on sites of incarceration; not as static physical entities, but as nodes in carceral networks which symbolise, represent and are experienced as, crystallisations of the penal philosophy of the prevailing state. In the context of hyperincarceration and the new punitiveness, it is argued that rather than being the primary loci of punishment and rehabilitation, prisons are now just some of the many nodes on the carceral continuum, in the context of a punitive state which operates in places far beyond the prison, through pervasive and pernicious policies which incarcerate and confine without actually imprisoning (Beckett and Murakawa 2012). Human geography is already in dialogue with notions of the punitive state, for example through concerns for the increased punitiveness of policies towards the urban homeless (De Verteuil et al. 2009, Laurenson and Collins 2007), zero tolerance policing (Swanson 2013), and political geographies of neoliberalism (Sparke 2006). In relation to incarceration, geographers such as Gilmore (2007), Peck (2003), Peck and Theodore (2009) and Shabazz (2009) have rightly drawn attention to the prison/ghetto symbiosis, focusing on the articulations of imprisonment with axes of urban disadvantage. A vibrant and significant avenue of research has already been identified in the work of Peck (2003) and Peck and Theodore (2009), along with Gilmore (1999, 2007) and Shabazz (2009), consolidated by Loyd et al. (2013), in relation to the 'institutions, practices and ideologies responsible for creating and maintaining a condition of mass incarceration' (Loyd et al. 2013: 7). Taking an abolitionist perspective, Loyd et al. (2013) consider imprisonment to be a form of state violence, arguing against the overreliance on incarceration which normalises imprisonment, with far reaching consequences for individuals, families, communities and societies.

In light of the vibrancy of this on-going scholarship, the remainder of this section of the book opens a space for carceral geography to direct renewed attention to the prison itself as a lens onto these wider processes, both in terms of the prison in its functional state, and its afterlife as the 'post-prison'. Gill et al.'s (forthcoming) advancement of carceral geography beyond place-based accounts of confinement argues for a heightened appreciation of the carceral *circuitry* of people, things, policies and punishments that enmesh apparently distinct carceral spaces, and the two chapters which follow therefore emphasise the potential of further research into the design of carceral spaces and the notion of carceral landscapes, which each offer opportunities to uncover the circulation of carceral ideas and practices both in time and in space, in spaces which, through their design and functionality, are material expressions of the relationship between the carceral and the state.

Firstly, Chapter 10 discusses the design and functionality of contemporary prisons, drawing attention to understandings of them as the physical manifestation of penal philosophies, arguing that carceral geographers could usefully turn their attention to prison design as a means of understanding what it is that prisons are intended to do, and the ways in which they achieve this through the deployment of space and architecture. Secondly, Chapter 11 takes forward the idea of a carceral cultural landscape, and the notion of the 'post-prison', no longer functioning as a space of incarceration, but nevertheless still imbued with, and able to communicate, important messages about the purpose of imprisonment both in terms of the of the system during which it was constructed, and during which it is protected, conserved, demolished, or left to decay.

Chapter 10
Prison Buildings and the Design
of Carceral Space

Prison design is crucial to the relationship between the 'carceral' and the state, in that it is the process which determines, in large part, how the goals of a criminal justice system are materially expressed. However, prison design remains under researched within criminology and prison sociology, and is yet to attract the attention of carceral geography. With this in mind, this chapter explores the significance of prison design, sketches out the extant research on this topic, and suggests areas of potential intersection between carceral geography, geographies of architecture, and health geographies, in the latter case specifically in relation to the notion of therapeutic landscapes.

Earlier chapters have explored the everyday experiences of carceral spaces in terms of space and agency, TimeSpace, and so on, and this chapter now takes forward the notion of prison buildings as coded, scripted entities which represent particular positions and imperatives, and the notion of the experience of these buildings as dynamic, multi-sensory, affective encounters. In so doing the chapter explores the design of prisons and the intentions behind their operation *as* prisons in terms of the imperatives of states and their criminal justice systems. It pursues the notion, developed within criminology and prison sociology, that the nature of carceral space has a significant role to play in understanding the extent to which the aims of a carceral system are translated into experiences of imprisonment. Beginning by briefly tracing the history and significance of prison design, with a focus on the UK, the chapter progresses through a consideration of architectural geographies' engagements with buildings as events and processes in relation to discourses of affect within human geography. It also draws together the notion of therapeutic landscapes, and extant research on prison design from criminology and environmental psychology, and finally discusses the utility of deploying these approaches for the study of prison buildings. It concludes by suggesting that the emotional or affective geographies of prisons as buildings, so far overlooked within carceral geography, could prove an important and significant avenue of future enquiry, in that these buildings can be read, and are experienced, as symbolic of the relationship between the carceral and a punitive state, in terms of who prisoners 'are', and what they represent, in the minds of those involved in producing buildings in which to imprison them.

Prison Design

As Wener (2012: 7) has argued,

> ... jails and prisons represent more than just warehouses of bed space for arrested
> or convicted men and women. They are more complicated environments than just
> good or bad, comfortable or not. The design of a jail or prison is critically related
> to the philosophy of the institution, or maybe even of the entire criminal justice
> system. It is the physical manifestation of a society's goals and approaches for
> dealing with arrested and/or convicted men and women, and it is a stage for
> acting out plans and programs for their addressing their future.

Prison design, then, is about more than accommodating and securing populations
from whom society needs to be 'protected' – although these two functions are
themselves challenging and complex. The design of a prison reflects the penal
philosophy of the prevailing social system, its ideas about what prison is 'for', and
what it is considered to 'do', and the messages about the purpose of imprisonment
that it wants to communicate both to prisoners, potential offenders, and to society
at large. As comparative criminology points out, offensive conduct is sanctioned
in different ways in different places. Punishment and crime are argued to have very
little to do with one another, with imprisonment rates 'to a great degree a function
of criminal justice and social policies that either encourage or discourage the use of
incarceration' (Aebi and Kuhn 2000: 66, cited in von Hofer 2003: 23) rather than
a function of the number of crimes which are committed. Imprisonment itself is
a sentencing decision – an outcome of an offence being committed, which in turn
depends on the definition of an offence, and the responses deemed appropriate for
different 'crimes' in different contexts. Imprisonment is not inevitable, therefore;
rather it is a conscious choice that societies, their governments and criminal
justice systems make, about the appropriate response to offending behaviour,
and the purpose of that response, in terms of the prevailing understanding of
what it is that prison is intended to achieve, both for society as a whole, and for
offenders themselves.

The following statement, made to *The Daily Telegraph* newspaper by Theresa
May, the then UK Home Secretary, neatly sums up the intentions of imprisonment
in the UK, as expressed by politicians to the electorate:

> Prison works but it must be made to work better. The key for members of the
> public is that they want criminals to be punished. They want them to be taken
> off the streets. They also want criminals who come out of prison to go straight.
> What our system is failing to do at the moment is to deliver that for the public.
> And that's what we want to do. (*The Daily Telegraph*, 15 December 2010)

The notion that 'prison works' is itself highly contentious (see for example Burnett
and Maruna 2004), begging the question about the ends which it is intended to

achieve. In her statement, the Home Secretary's three aims of imprisonment, to remove ('taking them off the streets'), to punish, and to rehabilitate ('going straight') characterise most prison systems, albeit the extent to which prison can achieve all, or indeed any, of these ends is highly debatable, and the balance *between* these aims as stated in public discourse, and as manifest and experienced in the criminal justice system itself, can vary widely. Just as prison design has yet to be foregrounded in academic literature, it also seems strangely largely disconnected from public discourses of imprisonment, despite being an integral part of prison commissioning and the expansion of the carceral estate.[1]

Whereas the United States and Western Europe are highly incarcerative (or perhaps hypercarcerative), other countries are by contrast *decarcerative*, actively deploying different techniques and sanctions to decrease their prison populations. This difference reflects a different underlying principle of imprisonment. For example, a 'less eligibility' principle informs much prison policy in the US and Western Europe, based on an understanding that prisoners should 'suffer' in prison, not only through the loss of freedom but also by virtue of prison conditions, which should be of a worse standard than those available to the poorest free workers. In other contexts, such as in Finland, prison conditions are intended to correspond as closely as possible to living conditions in society (Ministry of Justice of Finland, 1975), with the intention that penalties for offences are implemented in such a way that they do not unduly interfere with prisoners' participation in society, but as far as possible, promote it. The intention here is neither to oversimplify nor to romanticise the 'penal exceptionalism' of the Nordic countries (Pratt and Eriksson 2012, Ugelvik and Dullum 2012, Shammas 2014), but rather to point out that both the different philosophies of imprisonment and the different relative prison populations which these deliver, require and enable different intentions to be translated into the built form of prisons.

In the region of Anglophone penal 'excess' rather than Nordic 'exceptionalism', as the UK Home Secretary's comments suggest, prison must not only deliver a punished offender, but must do this in a way which satisfies the assumed punitiveness of the public; those whose apparent desire for 'prisoners to be punished' she is trying to satisfy. This notion of 'public punitiveness' has been discussed and contested in detail within criminology and prison sociology (Garland 2001, Greer and Jewkes 2005, Young 2003, Hancock 2004, Frost 2010); it is argued that growing public punitiveness reflects the profound anxiety that besets contemporary life in the 'risk society' of late modernity (Beck 1992, Bauman 2000), enabled by a media which heightens levels of crime consciousness, and correspondingly high levels of fear of crime. Despite recent research into the nature of punitiveness in public opinion, which has critiqued understandings of the relationship between a punitive public and increasingly punitive criminal justice policies (Frost 2010, Hamilton 2014),

---

1   In the UK, a former Home Secretary recalled that he was never asked to adjudicate on matters of prison design, rating 'the prison designs of much of the post-war period' as 'shoddy, expensive and just a little inhuman' (Hurd 2000: xiii–xiv).

and Ramirez' (2013: 329) observations that the limitations of data pertaining to what he called 'punitive sentiment' mean that actually very little is reliably known either about the opinions of given societies, or how these change over time and in relation to criminal justice policies, harsher sentencing policy is often attributed in part to demands for 'punishment' from what is assumed to be an increasingly punitive public, whose views therefore indirectly and directly affect criminal justice policy.

For the purpose of this chapter, however, the point is that in the UK, the context whose history of prison design is under discussion, prisons must today both punish and *be seen* to punish, as well as removing offenders from society in order to deliver some form of rehabilitation which reduces their future likelihood of reoffending. The fact that the UK prison system's reoffending rates remain stubbornly high, with 40 per cent of prisoners offending within 12 months of release, perhaps suggests that the punishment element of imprisonment (which also apparently fails to act as a deterrent to offending), outweighs its rehabilitative function. Although there is no transparent, linear translation of 'punishment' into prison design, as UK prison architecture has evolved, the interplay between philosophies of punishment and theories of prison design have resulted in preferred types of building thought capable of accomplishing the prevailing goals of imprisonment – which themselves have changed over time as penal philosophies have ebbed and flowed (Johnston 2000: 1).

A comprehensive survey of the history of UK prison design and the interrelationships between the various influences which have affected it (considered in detail by Brodie et al. 1999 and 2002, and Fairweather and McConville 2000), is beyond the scope of this chapter. However, considering prison buildings as scripted expressions of political-economic imperatives, the aesthetics of prison buildings are imbued with cultural symbolism (Moran and Jewkes forthcoming). Focusing on the UK carceral estate, although there is no 'typical' prison, for the majority, exterior architectural features render them instantly recognisable, within that cultural context, as places of detention and punishment. Mid-nineteenth century prisons, for example, were built to resemble fortified castles (e.g. HMP Leeds 1847), or gothic monasteries (e.g. Strangeways 1868), and exterior facades communicated the perils of offending and the retributive power of the state. The twentieth century gradually saw a more utilitarian style reject the decorative aesthetic, communicating an ideal of modern, 'rational' justice and authority (Hancock and Jewkes 2011). By the 1960s and 1970s, new prisons such as Gartree and Long Lartin, whilst still communicating authority and efficiency, echoed the austere, functional, styles of high, progressive modernism (ibid.). By the end of the twentieth century, UK prison architecture demanded higher walls, tighter perimeters and heightened surveillance, in response to earlier escapes, riots and security breaches,[2] and in parallel with the rise of 'new punitiveness' in wider

---

2   1990s security breaches included prisoner rooftop protests at Strangeways in 1990 and escapes from Whitemoor and Parkhurst prisons in 1994.

criminal justice policy. The evolution of prison architecture has at various points been intended to communicate a message about the nature of the imprisoning state and the legitimacy of its power to imprison, with the 'audience' for the various messages of this architecture being the inmate who receives the punishment handed down by the state, and society at large to whom imprisonment as punishment must be legitimated (Moran and Jewkes forthcoming).

The imprisonment of offenders today takes place within a framework of primary international covenants and conventions, such as the Convention Against Torture and Other Cruel, Inhuman or Degrading Treatment or Punishment, which are intended to guarantee proper treatment for those in detention under all circumstances. These instruments do not set explicit standards for the treatment of prisoners, but they provide a means of monitoring basic standards of humane treatment. Driven by a concern for the *treatment* of those detained, these conventions do not, therefore, make prescriptions about the exact nature of prison buildings, especially in terms of their outward appearance and architectural style.

Contemporary UK penal architecture reflects government reports commissioned within this policy context, which transformed prison security and with it prisoners' quality of life (Liebling 2002, Drake 2012). A preoccupation with 'hardening' the prison environment to design-out risk through environmental modification coincided with the UK prison service becoming an executive agency in 1993, and with the early 1990s enabling of private contracts for the design, construction, management and finance of penal institutions. An approach to prison control based on a balance between situational and social control has arguably swung towards an understanding of the situational dependence of behaviour, 'creating safe situations rather than creating safe individuals' (Wortley 2002: 4). In recent years UK prison new-builds have been driven by logics of cost, efficiency and security, and a need to comply with HM Prison Service Orders about the specification of prison accommodation, which lay out 'measurable standards' which can be 'applied consistently across the estate' in order to enable the prison service to provide 'decent living conditions for all prisoners' (HMPS 2001: 1). In this context, prison exteriors have tended to adopt a bland, presumably cheap, unassuming and uniform style with vast expanses of brick, few, small windows and no unnecessary decoration (Jewkes 2013a). Internally, the imperative in spending the Ministry of Justice's approximate £300m annual capital budget is to deploy indestructible materials to create custodial environments with no ligature points in which prisoners cannot physically harm themselves or others (RICS 2012). For example, one of the most recent UK prison new builds,[3] constructed as a part of the 'custodial architecture' portfolios of a specialist building contractor, was described as 'very operationally efficient' with 'a modern custodial aesthetic'. Advertising their 'Custodial and Emergency Services' project capabilities, the

---

3   David Nisbet, 'Partner at Pick Everard', press release, 11 June 2012, http://www. pickeverard.co.uk/news/2012/Pick-Everard-completes-UKs-largest-public-funded-prison-project.html, accessed 3/2/2014.

contractor, whilst conceding the need for a prison building to 'have a positive impact', and to be 'safe, non-threatening, secure and aesthetically pleasing', highlighted the imperative for 'value for money [to] be carefully balanced against the need for robustness and security'. Their experience and expertise in this area was described as bringing 'efficiencies at the design stage' including the kind of modified environments that create safe situations (Wortley 2002), such as 'designing ligature free environments by incorporating junctions and fixing details within structural walls and floors'.[4]

Nineteenth century prison buildings still in service are usually considered the least desirable environments within the UK penal estate. But while these Victorian 'houses of correction' ensured inmates' restricted economy of space, light and colour, imprisoning psychologically as well as physically, as argued elsewhere (Moran and Jewkes forthcoming) it has yet to be established empirically whether 'old' always means 'bad', while the kind of 'contemporary' prison described above necessarily equates to 'progressive' or 'humanitarian'. For example, within a year of re-opening in 1983, the 'new' Holloway Prison was criticised by the UK Prisons Inspectorate as engendering a form of torture that could result in acute mental illness. Levels of self-harm, suicide and distress were high and vandalism, barricading of cells, floodings, arson and violence against other prisoners and staff were common (Medlicott 2008). Among interior layouts recently advocated to manage problems like these is the 'new generation' campus-style arrangement of discrete housing units connected by outdoor space and flexible planning and design. Such prisons have experienced different levels of success; although prison architecture may reflect underlying penal philosophies, the ways in which it is experienced depend heavily on local contingencies and on the human subjectivity of the habitation of buildings. For example, Feltham and Lancaster Farms Young Offenders Institutions have been perceived differently on issues such as bullying, self-harm and suicide; Lancaster Farms has been held up as a 'shining example of commitment and care' (Leech 2005), whilst Feltham's reputation is coloured by years of damning reports and a high-profile murder (Jewkes and Johnston 2007). The differences between these two similarly designed institutions indicate that although, as Adey (2008: 440) argues, 'specific spatial structures … can … have certain effects', seductive spatiality (Rose et al. 2010) or ambient power (Allen 2006), which direct or shape human behaviour, this process is not at all straightforward.

Research into Prison Design

As early as the 1930s, architectural researchers pointed out the importance of prison design in shaping the experience of incarceration. In 1931, Robert Davison,

---

4   http://www.pickeverard.co.uk/custodial-emergency-services/index.html, accessed 3/2/2014.

former Director of Research for the *Architectural Record*, published a caustic article which castigated both US prison commissioners for lack of knowledge about what they wanted new prisons to achieve, and penologists for being 'surprisingly insensitive to the enormous importance of the building in the treatment of the prisoner' (1931: 39). Recognising that the design of prisons seemed to be a blind spot for the criminal justice system, he advocated that it was the job of the architect, even though they could 'scarcely be expected to be a penal expert', to indicate the 'necessity for a prolonged and careful study of this problem', and for a 'thorough research in [prison] building' (ibid.).

Despite the subsequent expansion of the penal estate and the immense investment in prison building in the UK and elsewhere, prison design has received remarkably little academic attention, and Davison's 'prolonged and careful study' is still to materialise. In the early 1960s, interest in new prison architecture and design reached its peak, with a special issue of *British Journal of Criminology* was devoted to the topic. In subsequent decades, however, criminological interest in this subject seems to have waned; scholarship on prison design has been sparse and its focus has been largely historical rather than contemporary, tracing the eighteenth- and nineteenth-century 'birth of the prison' (e.g. Johnston 2000). The dearth of scholarship on this topic is remarkable since the voices of prisoners, reflecting their experiences of incarceration in media such as autobiographies and poetry, speak vividly of prison design and its effects on the lived experience of incarceration (e.g. Hassine 2010, McWatters 2013). However, whilst criminological prison research has long been dominated by Sykes' (1958) notion of the 'pains of imprisonment', recent work has started to consider new and different ways of understanding the experience of incarceration, which lend themselves more readily to dialogue with the notions of carceral space and prison design. Encompassing discourses of legitimacy and non-legitimacy (Sparks et al. 1996); security (Drake 2012); therapy (Stevens 2012); compliance and neo-paternalism (Liebling with Arnold 2004, Crewe 2009); quality of life and healthy prisons (Liebling 2002, Liebling with Arnold 2004); normalisation (Jewkes 2002); the depth, weight and tightness of imprisonment (Crewe 2009); the resurgence of the doctrine of less eligibility (White 2008), and public acceptability (Liebling with Arnold 2004) these studies hint at, if not fully articulating themselves with, notions of prison design.

The late 1980s saw a fleeting interest in prison design and prisoner wellbeing emerge within environmental psychology, with research identifying a link between physical environment and social climate (Houston et al. 1988) and finding that prison architecture which creates overcrowded conditions causes significant stress to inmates (Schaeffer et al. 1988). Although Canter (1987: 227) argued that a 'systematic, scientific evaluation of the successes and failures' of prison design was urgently required in order to explore this relationship further, no such evaluation has taken place, and in the intervening period, research in environmental psychology has tended to focus its attention chiefly on negative prisoner behaviours and the risk factors which are perceived to contribute towards them; for example, focusing on 'hard' prevention techniques for prison suicide, such as developing cell designs with no ligature points from which prisoners can hang themselves. In other words, focus has shifted away from a concern for social climate, towards the designing-out of risk of physical harm from prisoners' destructive behaviour through environmental modification, and by maximising control on the part of the prison authorities (Tartaro 2003, Krames and Flett 2000). A recent attempt has been made to establish a broad-brush link between different architectural types in the US (as determined by satellite imagery) and the carceral regimes assumed to exist within them, and 'misconduct' on the part of inmates (Morris and Worrall 2010), although the methodology necessarily precludes further explication of the means by which any such linkages take form.

Despite guarded transdisciplinary recognition that the design of carceral spaces has a direct effect on prisoner behaviour and control (Foucault 1979, Alford 2000), the lived environment of prisons, including its potential for positive experience, has been relatively overlooked in recent scholarship. Moreover, the dominance of psychological methodologies in extant research on the prison environment has delivered rather a narrow range of largely quantitative studies, based on, for example: urine tests to determine stress responses (Schaeffer et al. 1988); the deployment of suicide or misconduct statistics as a proxy for stress, towards which the physical environment might (or might not) be a contributory factor (Tartaro 2003, Morris and Worrall 2010); and true/false questionnaire responses as part of the Correctional Institution Environment Scale (CIES), which lacks an explicit environmental dimension; simply being used to measure 'wellbeing' in different institutions (Houston et al. 1988). At the other end of the methodological spectrum, in his work with prisoner poetry, McWatters adds to understandings of how prison space is actually experienced by those for whom 'it is an ordinary space of daily life' (2013: 199), describing carceral space as 'more plastic, fluid and manifold than totalizing notions permit' (ibid. 200), and arguing in support of efforts to expand the imaginary of lived spaces of incarceration.

Having recognised that the carceral environment 'matters' to prisoners' experiences, and having demonstrated it to some degree using a variety of methodologies, without exception, these studies call for a more nuanced investigation of the impact of design on those using and occupying prison spaces.

The 'new punitiveness' discussed earlier in relation to the relationship between the carceral and the state, comes clearly into view in relation to the prison estate, expanded to accommodate those imprisoned under circumstances of increasing prison sentences, increasing prison sanctions, and more austere and Spartan prison conditions, operating to a greater or lesser extent in various contexts (Pratt et al. 2013, Hallsworth and Lea 2011, Lynch 2011, Snacken 2010). This recent 'hardening' of penal sensibilities, which in the UK and elsewhere is coupled with more severe sentencing policies (Criminal Justice Act 2003), the fetishizing of risk and security within and outside the penal estate, and a rising prison population (which, in England and Wales, grew by 30 per cent since 2001, peaking at 86,842 in September 2011), makes questions of prison design and the lived experience of carceral space particularly pertinent. Although chronic overcrowding, high rates of drug use, mental illness, self-harm and suicide, recidivism and its associated financial and social costs, mar the UK system, prison escapes have fallen dramatically, due in part to prison design; prison walls are higher, prison space is sequestered through zoning, and CCTV cameras and other technologies proliferate. In the United States, Morin (2013: 381) argued that one contemporary facility is an example of the 'latest punitive phase' in American penology, one that neither simply eliminates, as in the pre-modern spectacle, nor creates the docile, rehabilitated bodies of the modern Panopticon; rather, she argued that this particular prison design is a 'late-modern structure that produces only fear, terror, violence, and death'.

Architectural Geographies, Therapeutic Landscapes and Prison Design

Human geography has over the past 10 years seen affect and emotion re-emerge as a major theme, with a growing body of work both *in* embodied, emotional and affectual geographies, and discussing the relationships *between* them (e.g. Anderson and Smith 2001, McCormack 2003, Thrift 2004, Bondi 2005, Pile 2010). Drawing on this scholarship, Rose et al. (2010) argued that emotional and affectual geographies have percolated into geographies of 'big things' (Jacobs 2006) such as airports, tower blocks, office blocks, shopping malls, libraries and ships, in that geographers interested in such 'big things' have begun to reflect this distinction between emotion and affect. Some explore '"feelings" in terms of the emotions expressed by human subjects, while others work with a version of affect that creates bodily behaviour and sensory perception with little or no mediation by subjective processes' (ibid. 339). Prisons might also be labelled 'big things', but as yet there has been little attention paid to the emotional or affective geographies of experiences within or about them, whether referring to prisoners, visitors, or, for that matter, any of the myriad people who engage with the carceral.

Recent work within carceral geography has addressed the significance of carceral space (Moran et al. 2013a), recognising space as more than the surface where social practices take place (Gregory and Urry 1985, Lefebvre 1991, Massey 1994), but although geographers understand that space can affect the ways people act within it, and are increasingly applying this perspective to carceral spaces, Siserman (2012) points out that studies of prisons as buildings and environments where the behaviour of inmates can be dramatically changed, and which investigate how this might happen, remain scarce.

Recent commentaries within architectural geographies and cultural geographies of buildings (e.g. Kraftl 2010, Jacobs and Merriman 2011, Rose et al. 2010, Jacobs 2006, Kraftl and Adey 2008) have argued for the importance of considering buildings in a number of connected ways; as everyday spaces in which people spend a significant proportion of their lives, as expressions of political-economic imperatives that code them with 'signs, symbols and referents for dominant socio-cultural discourses or moralities' (Kraftl 2010: 402), and in terms of perspectives that emphasise materiality and affect.

Although a representationalist focus on prison buildings as sites of meaning, symbolic of intentions and imperatives is itself arguably underdeveloped in prison scholarship, carceral geographers could go beyond the symbolic meaning of prison buildings, to consider the 'inhabitation' of prison buildings (Jacobs and Merriman 2011: 213), in terms of the 'dynamic encounters between buildings, their constituent elements and spaces, inhabitants, visitors, design, ergonomics, workers, planners, cleaners, technicians, materials, performances, events, emotions, affects and more'. Like any other buildings, prisons are sites in which a myriad of users and things come into contact with one another in numerous complex, planned, spontaneous and unexpected ways, and where the encounters are both embodied and multi-sensory (haptic, visual, acoustic, kinaesthetic, thermal and so on), and resonant of the power structures which exist both within and outwith the prison building and which shape its inhabitation. By seeing prison buildings in this way, carceral geography has the opportunity to respond to Rose et al.'s (2010) appeal for human subjectivity to be foregrounded within geographies of architecture, and to go some way towards exploring the 'nitty-gritty, material, localised details of architectural design and form-making' (Kraftl and Adey 2008: 228) in carceral contexts.

Where prison design is considered to impinge upon the lives of those inhabiting carceral space, the 'potential capacities' which prisons have 'to affect their inhabitants in certain ways' (Kraftl and Adey 2008: 228) are almost always considered to be harmful rather than therapeutic, as recently demonstrated by Raphael Sperry's campaign in the US for architects to boycott the design of Supermax facilities. In 2005, Architects/Designers/Planners for Social Responsibility (ADPSR) launched a 'Prison Design Boycott' urging architects to decline death chamber and Supermax prison commissions on the grounds of human rights violations – Supermax prisons being considered to inflict torture through long-term solitary isolation. Sperry asked the American Institute of Architects to amend its Code of Ethics and Professional Conduct to prohibit the design of execution chambers and solitary confinement, strengthening the existing Ethics Code which calls for support of human rights by identifying the relevant United Nations standards and stating that AIA members should not be involved in the designing of buildings intended to accommodate specific activities that violate human rights.

In the light of this campaign, considering prisons as potentially 'therapeutic' landscapes may appear obtuse, at the very least. However, recalling the earlier discussion of the purposes of imprisonment, although it is perhaps easier to consider prison design in relation to the function of punishment, attention should also be drawn to its stated function of rehabilitation, and therefore to a discussion of the extent to which prison buildings can perhaps 'heal' as well as harm.

The notion of 'therapeutic landscapes', first introduced by Gesler (1992) and explored further within geographies of health and care (e.g. Milligan et al. 2004, Gesler et al. 2004, Williams 2002, Laws 2009, Curtis et al. 2007), suggests that certain environments promote mental and physical wellbeing, and that these landscapes can be 'created' as well as 'natural' (Milligan et al. 2004: 1783). Based on an understanding of the ways in which environmental, societal and individual factors intersect to promote healing and wellbeing, 'therapeutic landscapes' reflect a sense of place as relational, and a holistic model of 'health' that encompasses the physical, emotional, spiritual, societal and environmental. Milligan et al. (2004) noted, though, that discourses around therapeutic landscapes have tended to focus on famous or one/off events or places, rather than on everyday, quotidian spaces, and to highlight the health-promoting aspects of historical places, rather than to illustrate how it might be possible to 'develop everyday places that promote ... physical and mental wellbeing' (ibid. 1783). Although prisons may not be quotidian spaces as Milligan et al. (2004) understood them, for prisoners and prison staff they are indeed the spaces of their everyday lives.

Although the therapeutic landscape approach has only recently been applied to prison buildings (see Moran and Jewkes 2014), other institutional spaces have been considered in this light. For example, Gesler et al. (2004: 117) discussed the debate in the UK over what constitutes 'good hospital design' in conditions in which expert discourses of costs and clinical functionality tend to take precedence, but in which there is an acceptance that hospital buildings need to 'work' to promote patient recovery and healing, in ways which resonate with a rehabilitative interpretation of imprisonment. Their conception of hospital buildings as therapeutic environments in terms of physical, social and symbolic space has much in common with Kraftl and Adey's (2008) suggestion that buildings can engender certain affects and evoke certain types of situations.

Recent developments both in prison architecture and design (albeit outwith the context of the 'new punitiveness'), and in criminological research into prison aesthetics and 'anaesthetics' (Jewkes 2013a), echo Kraftl and Adey's (2008: 228) suggestion that one function of buildings can be an attempt to stabilise affect, 'to generate the possibility of pre-circumscribed situations, and to engender certain forms of practice, through the design and planning of buildings, including aspects such as form and atmosphere'. In their own work, they found that certain generic expressions of affect evoked certain kinds of inhabitation, materialised via buildings in their 'potential capacities to affect their inhabitants in certain ways' (ibid.). In other parts of the world, in which the 'new punitiveness' of the US, UK and elsewhere has not taken hold, prison designers have focused on the rehabilitative function of imprisonment, and have experimented with progressive and highly stylised forms of penal architecture, with internal prison spaces with soft furnishings, colour zoning, maximum exploitation of natural light, displays of art and sculpture, and views of nature through vista windows without bars.

For example, in designing a planned women's prison in Iceland, the project team from OOIIO Architecture[5] intended 'to design a prison that doesn't look like a prison, forgetting about dark spaces, small cells, and ugly grey concrete walls … we based the building design on natural light, open spaces, and natural green materials like peat, grass and flowers'. Instead of designing one large building (like a 'typical repressive old prison'), they decided to break it into several 'human-scale, connected' pavilions, which must be efficient and functional to enable the spatial separation of prisoners, but which must have 'natural light and exterior views, to increase the feeling of freedom'. The architects also had an eye to the speed and ease of construction, and to the eco standards of the building, planning to draw upon Icelandic vernacular architecture to insulate the building. With a facade constructed from peat-filled cages, planted with local flowers and grasses, they intended to deliver a building 'that changes with the seasons', making prison life 'less monotonous and more human and natural related'.

This kind of design of new prisons, in Norway, Iceland and Denmark, arguably plays up and enhances certain generic expressions of affect connected to openness, flexibility and 'humane' treatment, to evoke certain kinds of inhabitation encourages personal and intellectual creativity, and even a lightness and vividness of experience (Hancock and Jewkes 2011).

Directions: The Design of Prison Buildings and their Affectual Potential

Kraftl and Adey (2008: 228) called for further research into the ways in which architectural forms try to manipulate and create possibilities, and into how those affects are experienced and negotiated in practice, via the notion of inhabitation. This chapter concludes by suggesting two avenues of potential future research in carceral geography, building on Kraftl and Adey's (2008) work; first, paying attention to the processes of architectural design and construction in order to uncover and question the multiple political, affective and material ways in which prison buildings are designed and constituted, and second, attending to the ways in which the affectual potentialities of prison buildings are negotiated in and through practices of inhabitation.

If carceral geography is to explore the intentions behind the architecture, design and technologies of spatial management and control that characterise the penal estate, Morin's (2013) recent work highlights the value of such an approach, focussing on the US penitentiary at Lewisburg, Pennsylvania, which was retro-fitted in 2008 to offer the country's first federal Special Management Unit (SMU) programme of its kind. Designed for the most intractably troublesome federal inmates from around the country, the SMU features double-celling of inmates in tiny spaces, subject to 23-hour or 24-hour-a-day lockdown. These spatial tactics,

---

5   http://plusmood.com/2012/06/female-prison-in-iceland-ooiio-architecture/, accessed 3/2/2014.

she suggested, and the philosophy of punishment underlying them, contrast starkly with the modern reform ideals upon which the prison was originally designed and built in 1932. However, it is not just in the Supermax that such descriptions are used to describe everyday life in confinement. One 'lifer' who spent 27 years in the American penal system, before taking his own life, asserted that prison designers and managers have developed a 'precise and universal alphabet of fear' (Hassine 2010: 7), which creates an outwardly benign illusion but is comparable to an ant farm in which the visible order, regularity, and routine fails to expose the 'violence and crushing hopelessness the trapped ants are actually forced to endure' (ibid. 122). Such observations beg questions about the design process which, as Wener (2012: 7) noted, is 'the wedge that forces the system to think through its approach and review, restate, or redevelop its philosophy of criminal justice'. What are the processes which lead to the conscious and intentional design of such carceral spaces, and to what extent do prisoners experience in them what was intended in that design process? (Moran and Jewkes forthcoming).

In relation to the affectual and emotional geographies of prison buildings, carceral geography is well placed to deploy a holistic, multidisciplinary approach that considers the relationship between space, meaning and power, and the ways in which architecture and design communicate the aims and techniques of penal authority, shape the lived experience of imprisonment, and impact on the working environment of prison staff as well as the lives of prisoners. Future research could address the penal philosophies and imperatives underpinning the design of newly commissioned and newly built facilities, and explore the impact of prison design on the experience of imprisonment, on the behaviour of those who occupy and move through carceral spaces, and on staff-prisoner and staff-management relationships.

In this context, the 'dynamic encounters' (Jacobs and Merriman 2011: 213) which occur between the inhabitants of prison buildings, technologies operational within them, and the buildings themselves, are critical to understanding their inhabitation. The subjectivity and emotion of human experience mediates the operation of affect in all situations, but arguably in prisons its importance is heightened. Criminologists recognise that prisoners constantly 'manage' issues of self and identity and adapt socially under intense and inescapable duress, and the kinds of encounters between prisoners and prison staff which are encouraged by, or which are even possible within, differently designed prison buildings are worthy of investigation. 'Dynamic security' in which prison staff are encouraged to develop good relationships with prisoners through direct contact and conversation, is no longer possible in many new build prisons where staff are physically separated from prisoners. Where surveillance technologies enhance the observation of carceral space, some prisoners may value CCTV as a means of protecting their personal safety and for its capacity to provide evidence of bullying and assaults, but technologies also reinforce the absence of privacy and create additional stresses, for both prisoners and staff. The utilisation of surveillance and monitoring technologies in prisons as workplaces have inevitably brought prison employees under closer scrutiny from their managers (Townsend and Bennett 2003, Ball

2010), and it is argued that the notion of trust, once regarded as essential to prison management–staff relationships, has been undermined by surveillance systems which ensure that 'correct' organisational procedures are followed. Increasingly, prisons routinely monitor everyone passing through them via an interface of technology and corporeality, encouraging flexibility of movement while retaining high levels of security. Cameras wirelessly transmit digital images which are then screened for unusual objects and atypical movements; biometric and electronic monitoring of prisoners and visitors allows the tracking of bodies within in the prison; listening devices monitor the spectral content of sound to spot illicit use of mobile phones or early signs of aggressive behaviour; and prison officers' Blackberry-style devices enable immediate reports to be relayed to Security (OIS 2008) (Moran and Jewkes forthcoming).

Whereas Morin's (2013) work suggested that like the UK, the US is experiencing a trend towards increasingly severe and restrictive prison designs, elsewhere, prison buildings are being designed with different intentions in terms of the manipulation and creation of possibilities. In northwest Europe, decarcerative policies deliver smaller numbers of prisoners, and for these smaller prison populations, the use of surveillance technologies facilitate 'humane', open-plan, 'progressive' prisons with a greater degree of movement among and between inmates and staff, and a wider range of possible encounters. Even here, however, the appearance of these buildings, in terms of their natural materials, large windows and natural light, although conveying a sense of ease and relaxation, arguably replicate and perhaps enhance some of the issues of privacy, identity management and presentation of self-identified in more obviously 'restrictive' settings. Shammas (2014: 104) has called for attention to be paid to the 'pains of freedom' inherent in Norway's more 'humane' prisons. There is some evidence that technology-assisted, decentralised, podular designs approximate 'normality' by providing safer and more comfortable living environments, and removing security gates, bars and grilles, enabling prison officers to be more than 'turn-keys' (Spens 1994). But as Hancock and Jewkes (2011) have argued, there has been scant official or scholarly discussion of other potential uses of technology, such as the identification of abuse or aggressive behaviour by prison officers (either to prisoners or their colleagues), the surveillance of staff smuggling contraband into the prison, or behaving in ways disapproved of by prison authorities. Similarly, there is little debate about the moral and ethical implications of near-constant surveillance of prisoners and officers, or the difficulties in establishing trust when basic standards of privacy are compromised. The use of technologies could exacerbate complex horizontal and vertical relationships between prison inmates, officers, managers and ministers. Everyone who moves within and through these 'hyper-organisational spaces' (Zhang et al. 2008) is not only enmeshed in a surveillance assemblage that forces them to manage their own presentation of self within the regulative framework of the institution, but is further encouraged to watch while knowingly being watched. Although lack of privacy has long been recognised as a 'pain of imprisonment' for inmates, for prison staff the new Panopticism is a novel form of control (Hancock and Jewkes 2011).

With Wener's (2012: 7) proposition that 'the design of a jail or prison is critically related to the philosophy of the institution, or maybe even of the entire criminal justice system' in mind, this chapter has argued that an understanding of prison design could enable better understanding of the lived experience of carceral spaces and offer a starting point for future research in carceral geography. It also suggested that geographies of affect and embodiment might offer a useful theoretical and methodological lens through which to access the experience of the 'carceral', by shedding light both on how people experience the carceral, broadly defined, in terms of the feel of it, their feelings in it, and their feelings about it, in relation to the importance of affectivity (McCormack 2008, Merriman 2010).

As the first section of this book demonstrated, the lived experience of carceral spaces has emerged as a central theme of recent research, and carceral geographers have made valuable contributions to understandings of how, even within the most restrictive conditions of confinement, prisoners employ effective spatial tactics within surveilled space, create individual and collective means of resistance to carceral regimes, and succeed in appropriating and personalising carceral spaces (Moran and Jewkes forthcoming). While this growing body of work has illuminated some of the darkest carceral spaces, the majority of research to date has tended to focus on inmate responses to, and adaptations of, the physical spaces of incarceration, rather than to draw attention to the processes which led those spaces to *be* as they are, and what this means for the ends prison buildings serve for the state which creates them. Missing from this work is a consideration of the ways in which punitive philosophies are manifest in prison commissioning and construction, and thence in prison buildings themselves.

The challenge for carceral geographers, alongside scholars interested in prisons and imprisonment in other disciplines, is to start to address why those spaces are as they are, and to interrogate the intentions behind their design. Returning to Davison's (1931) condemnation of US prison design, research needs to illuminate the commissioning process, to uncover what it is that architects are asked to deliver, and how those demands are framed. Davison argued that prison authorities would 'never get the most out of their architects until specifications are presented not in terms of definite plans and materials, but in terms of performance' (1931: 33). He called for commissioners not to request cell blocks, but sleeping places; not to demand mechanical ventilation, but instead to require good air for every prisoner. Then, he concluded, 'let the solution be worked out. In many instances the result will be astonishing. It will not resemble the present jail at all' (ibid. 34). Designing a prison based on the requirements of the building, rather than simply accepting and replicating what has been built before, was for him the key to delivering 'better' prisons.

Pursuing these questions could enable us not only to better understand the experience of incarceration, but also to open the design process itself to scrutiny and reflection (Moran and Jewkes forthcoming). Wener (2012: 7) argued that prison environments represent both an 'overt' agenda that provides measurable quantities of space for accommodation, training, therapy, education and so on, but also a 'covert' agenda that reflects what or who inmates 'are' in the minds of planners, designers, and those who commission them to design and build prisons. By opening a space for the articulation of this 'covert' agenda, carceral geography, in dialogue with criminology and prison sociology, could contribute positively to the on-going debate over the expansion of the penal estate. In so doing, carceral geography could speak back both to architectural geographies and, in relation to therapeutic landscapes, to geographies of health, from study of a type of building as yet unexplored in either of these subdisciplines of human geography.

Chapter 11

Carceral Cultural Landscapes, Post-Prisons and the Spectacle of Punishment

This chapter explores the cultural geographies of a variety of carceral landscapes and their relationship with a shifting punitive state. Whilst the previous chapter focused on the design of prisons and the intentions behind their operation *as* prisons in terms of the intentions of a particular criminal justice system, in this chapter the focus is on the relationship between the carceral and a punitive state in a wider and less tangible sense. The chapter looks at prisons and places of incarceration as landscapes with meaning and power, and traces some of the ways in which what happens to prison sites when they cease to be places of incarceration (the notion of the 'post-prison') can and should remain deeply significant for their relationship to a punitive state.

After introducing the notion of prisons as carceral cultural landscapes, the chapter investigates the operation, preservation, conversion and re-use of prisons, the meanings which are embedded and experienced in these sites as cultural landscapes, and the spectacle of punishment as fetishized through the preservation and presentation of places of incarceration either as tourist sites (via 'dark tourism'), or as consciously converted and repurposed locales. It discusses the tensions inherent in memorialisation and commemoration of former prison sites in relation to contemporary contextual political tensions.

The chapter next engages with the notion of prison towns and prison zones as landscapes in which carcerality has become embedded and sedimented through generations of operation (for example in former Gulag regions of Russia), and the ways in which some such penal zones have themselves become virtual carceral landscapes through online resources and representations. It also explores media representations of carceral spaces which call for a purposeful engagement with virtual landscapes of incarceration, and require personal reflections on the purposes of imprisonment.

Carceral Cultural Landscapes

Viewing spaces of imprisonment as carceral cultural landscapes draws on decades of work in human geography. The 1980s notion of landscape as a 'way of seeing', or as a way of imagining the world (Winchester et al. 2003) is based on the notion that all landscapes are representations. As Robertson and Richards (2003) argued, the landscape and those who occupy it shape each other reflexively, by means of cultivation and building, and by the projection of aspirations and fantasies. This symbolic approach to landscape constructs it both as a cultural product and a cultural process, with social, cultural and political elites using their positions of power to promote their own values through inscription in the landscape. Landscapes can, therefore, represent the relationships of power and control out of which they have emerged, both in terms of the dominant ideologies, and those elements of resistance or alternative cultures whose values may also be inscribed into them. 'Reading' these landscape texts can mean using iconography in order to elucidate the meanings of the visual landscape (Daniels and Cosgrove 1988), or approaching the landscape via intertextuality (Duncan 1990, Duncan and Duncan 1988), drawn from literary theory, which posits that the 'reader' is as involved as the 'writer' in a landscape's construction and its decoding.

As understandings of landscape have shifted in balance to favour its interpretation as a fluid process rather than a static product, landscapes have increasingly been viewed as active and dynamic, and as a critical part of the processes through which identities are formed (Mitchell 1994). Moving beyond the 'ways of seeing' approach to landscape studies, more recently, scholarship in human geography has sought to incorporate notions of materiality, embodied practice and perception, and the intertwinings of self and landscape through a focus on their performance via everyday practices such as walking and visualising. For example, Wylie (2007: 69–72) presented landscape less as a series of layers to be excavated to uncover meanings and systems of power, than a texture to be searched horizontally.

Spaces of incarceration, as places through which prevailing criminal justice systems impose punishments on those deemed to have offended against the rule of law, are explicitly intended to promote the values of the state, and its dominant ideologies of justice and punitivity. As 'texts' which can be 'read', they form 'palimpsests' of identity and culture, which both validate and authenticate consensual notions of the justice, whilst simultaneously inviting alternative readings. They are also actively intended to act as fluid processes of identity formation, in that the function of these facilities, in terms of the balance between retribution and rehabilitation, sets out to have an impact on those incarcerated within them – itself dependent on the philosophy of imprisonment which has informed the construction of the prison in the first place, and the perceived purpose of imprisonment itself. The experience of imprisonment, of course, can also impact on individuals and their identities in myriad ways far beyond those intended by criminal justice systems.

Prison buildings as the cornerstones of carceral cultural landscapes, are certainly meant to mean *something*. Their exterior design has historically been intended to convey a certain sense of justice, a notion of the power of the state, and its ability to extract retribution for crimes committed. As Brown and Barton (2013: 2) noted, 'the powerful symbolic external structure of the prison [has been] built to intimidate and deter'. However, these meanings are contested, and prison sites can also become stages for tension between dominant narratives (of justice, and the power of the state) and alternative perspectives. As McAtackney (2013) has observed, prisons can symbolise both the power and the vulnerability of the state. For example, Robben Island prison, offshore of Cape Town, South Africa, the place of incarceration of anti-apartheid campaigner Nelson Mandela, was intended as a symbol of the power of the apartheid state against the pressure for democracy, but also became a symbol of racial oppression and a crucible for human rights protest, and the focus for appeals for Mandela's eventual release.

The site of Long Kesh/The Maze Prison, Northern Ireland, is another example of a carceral cultural landscape with deep political and social significance. Used as a place of incarceration for both Republican and Loyalist paramilitaries during the 'Troubles', on the one hand it stood as a symbol of the legal process, but on the other was viewed as a symbol of oppression of Republicans, and became a focus for political unrest and demonstrations. As a site of embodied political practice and perception, Long Kesh/The Maze can be read as an assemblage of close intertwinings of self and landscape, on the part of 1980s Republican hunger strikers, and decades later, in terms of the discourses around its closure and repurposing. Initially a vacant World War II airfield pressed into service as a temporary internment camp in the early 1970s, prisoners at Long Kesh/The Maze had previously been accommodated in disused Nissen huts, until eight new purpose-built H-shaped 'H Blocks' blocks were added to the site between 1975 and 1978 (McAtackney 2005, 2013). These prominent features of the prison landscapes gave their name to the 'H-Block Campaign' which accompanied the 1980s hunger strikes at Long Kesh/The Maze, centring on demands by Republican prisoners to be viewed as political prisoners. The hunger strikes inside Long Kesh/The Maze were supported by demonstrations, protests, riots and work stoppages outside the prison, and the election of two hunger strikers as 'Anti-H Block' candidates to the British Parliament and the Irish Dail. The carceral landscape of Long Kesh/The Maze became a place of direct contestation of the interpretation of validity of British rule of Northern Ireland. By 1979, five demands had been framed under the 'Smash H-Block' campaign, organised by the National H-Block/Armagh Committee. Prisoners demanded exemption from wearing prison clothes, and from prison work, freedom of association with fellow political prisoners, the right to educational and recreational facilities and visiting, and entitlement to full remission of sentences. In other words, their contestation of the validity of government took form through claims against the deployment of the space of the prison, and the specific techniques of day-to-day management of the prison and the prisoners within it. Hunger strikes, in which the frailty of the body was instrumentally exposed as a means to draw attention to perceived injustices, therefore became a prominent and highly publicised expression of resistance to both the carceral regime and the form of government it represented.

Post-prisons and the Repurposing of Prison Buildings

Carceral systems are constantly evolving, in terms of sentencing policy, demands placed on the carceral estate, and overarching views of the purpose of imprisonment. Phases of prison building and prison closure mean that prisons eventually come to the end of their functional life, raising questions about the fate of the buildings themselves, and their inscription and interpretation as cultural landscapes. What happens when prisons cease to incarcerate depends in part on their context and circumstances, the uses to which the buildings can practically be put, and the political, social and cultural significance of the sites themselves. Whether a 'post-prison' existence, (in which a post-functional prison building takes on a new usage completely disconnected from its previous function), is even possible, is debatable. It seems that whatever the purpose (if any) to which a prison site is subsequently put, vestiges of its previous function remain inscribed in its cultural landscape, whether these are uncomfortable reminders, commercially attractive qualities, or a combination of the two. As Vanderburgh (1992) suggested, various attitudes towards the past with which prisons may be associated, (such as a critical attitude in which events are edited out, or an antiquarian attitude, which seeks to preserve), affect the fates of those prison sites.

Simply as pieces of physical infrastructure, large, sturdy decommissioned prison buildings, if located in commercially viable places, have various potentially profitable uses, such as conversion into housing or hotels, as in the instance of the Hotel Lloyd in Amsterdam (Ong et al. 2012). In such cases, even if the former function of the site is incidental to its commercial appeal, arguably the post-prison still remains a prison, and its identity becomes part of the narrative of the repurposed site. For example, at the redeployed sites of Oxford Castle prison, UK, Katajanokka prison, Helsinki, and at Langholmen prison, Stockholm (now hotels) (see figures 11.1, 11.2 and 11.3), carceral history and architectural style are consciously marketed selling points; some of the sites have 'interpretation centres' to tell a potted history of the site to paying guests, and some sell prison-themed mementoes.

Prisons have also been converted into retail premises like the shopping mall created out of the fabric of Punta Carretas prison in Montevideo, Uruguay, and in the conversion of these sites, both the physical and the cultural legacy of the prison has to be addressed.

Opened in 1910 as the exemplar of model prison architecture in Uruguay, Punta Carretas prison's aim was to rehabilitate individuals through humanitarian punishment, and its opening, which coincided with the abolition of the death penalty, was characteristic of the Uruguayan state's modernization plans (Ruetalo 2008, Draper 2012). However, by the 1930s Punta Carretas was holding political prisoners, becoming by the 1970s Uruguay's most important centre of confinement for politicals, until a mass escape in 1971 saw the prison nearly emptied, and the remaining prisoners moved to the new military *Libertad* Penitentiary. Post-1970s, the prison held mainstream prisoners, and eventually it was slated for closure, a

Figure 11.1   Merchandizing penal history at Katajanokka Hotel, Helsinki

Figure 11.2   Merchandizing penal history at Langholmen Hotel, Stockholm

Figure 11.3   Utilising carceral symbolism at Langholmen Hotel, Stockholm

decision initially triggered by rising local property values, but then delayed by the recognition that Punta Carretas was a national site of cultural heritage, which should be preserved.

With the cost of the preservation of the building proving too high for the then post-dictatorship state, a decision was taken to preserve the building, but to put it to commercial use. The prison was converted into a shopping mall in 1994, as part of a wider process which Draper (2012: 23) described as a means to 'envision the country of the future – that is, the country of consumer services'. In converting the prison, a conscious effort was made to selectively demolish and preserve certain features, in an effort to 'preserve the spirit of the prison, but in a way in which this 'preservation' would not be an obstacle to developing its new function' (architect Estela Porada, cited in Draper 2012: 48). Draper drew particular attention to this notion of the 'spirit' of the prison, suggesting that the spirit is connected to 'leaving behind the prison in architectural form without bringing forth the painful, past spectres of this site' (ibid. 49).

Discussing the actual space of the prison-mall, Draper described the preserved relics of the prison – the façade, the gateway through which prisoners were previously led to their cells, the walkways on former prison landings, and the former cells which now contain shops, food courts and entertainment complexes. She particularly focused on the tensions in the cultural landscape between the inside and outside of the mall, specific relics of the prison which appear disconnected from the mall itself, and the effect of the disguised and fetishized remnants of the prison within the colourful mall. For carceral geographers, Draper's work, which also analyses the literary afterlives of Punta Carretas (writings on the prison itself and its transformation), offers an intriguing reading of this site, weaving together ideas of spirits and spectres, idioms and residues, evocation and translation, and the relationship between the unique and the universal.

More transient, 'stop-gap' repurposing of prisons still plays on the marking out of carceral landscapes, and includes the use of prison sites as film sets or backdrops, such as the exterior of the disused St Albans Prison, UK, used in the opening credits of the British TV sitcom 'Porridge'. However, after closure some prison buildings simply fall into decay and disrepair, left to weather and decay in ways which, ironically, may increase their future commercial value. In many cases this stasis emerges from a lack of viable options for profitable conversion, perhaps based on the nature of the prison buildings, or where they happen to be, as in the case of Joliet prison, Illinois, USA, in a low-cost area, away from potential commercial users, with little potential for conversion, coming to the end of its functional life in economic circumstances which preclude investment for conversion. In other instances, such as Patarei in Tallinn, Estonia (Kuusi 2008), this former Soviet prison in a 'prime' seafront position has been acquired for redevelopment, but investment has not been forthcoming, and the site is being allowed to decay, with low-level marketing as a macabre visitor attraction, and intermittent use as an edgy music venue, bar, and alternative site for art installations (figures 11.4 and 11.5).

In other cases, stasis rests on the perceived controversy of the specific site and a range of sensitivities around the future uses to which it might be put. Debate often focuses on whether a prison site should be preserved at all, with arguments for conservation driven by a combination of concerns for the architectural importance of the buildings themselves as heritage objects, and/or the history of the site as a place of imprisonment, perhaps in relation to significant or prominent former inmates. Because former prison sites can be so important and potentially sensitive, reaching agreement about how they should be preserved can be a long and tortuous process. Long Kesh/The Maze, for example, is now recognised as one of the key heritage sites of the Northern Ireland conflict, within the context of the political processes which followed the Belfast Peace Agreement in 1998. Following its closure in 2001, a 2005 'Maze Consultation Panel' proposed the reconstruction of part of the site as an International Centre for Conflict Transformation. Although various stakeholders responded to the consultation, only Irish Republican political party Sinn Féin articulated a clearly defined sense of the heritage value of the site, and an understanding of how it might be appropriated and exploited as an iconic place for remembering, contestation and resistance. Sinn Féin's favoured type of preservation resembled Robben Island or Auschwitz-Birkenau, but this approach was considered highly politically sensitive, inscribing a particular and partial reading of Northern Ireland's history at the site, and in the opposing view of Unionists, risked creating a shrine to the 'armed struggle' and the hunger strikers (Neill 2006, Graham and McDowell 2007). Today, the majority of the structures

Figure 11.4   Patarei prison, Tallinn, Estonia

Figure 11.5   Art installation at Patarei prison, Tallinn, Estonia

on site have been demolished, leaving a representative sample for interpretation, and plans for an EU-funded Peace Building and Conflict Resolution Centre have recently been passed (McAtackney 2013). The site remains closed to the public and very delicate negotiations over its future use and the nature of the site are ongoing.

Prison Heritage Sites

The preservation and presentation of places of incarceration as prison heritage sites, or as converted and repurposed locales which consciously rehearse and reinterpret a carceral past, speak to wider discourses around 'dark heritage' and 'dark tourism' in general (Foley and Lennon 2000), and specifically to the notion of prisons as sites of 'dark heritage'. The concept of 'dark heritage' pertains to consumerist interest in 'dark sites' of death and disaster, and the conscious preservation and marketing of such sites in ways which enhance their macabre appeal.

The consideration of places of incarceration as 'dark sites' has been explored in research in a range of locations (Stone 2006), for example by Shackley (2001), Strange and Kempa (2003), Deacon (2004), and Phawana-Mafuya and Haydam (2005) in Robben Island, South Africa; Walby and Piché (2011) in Canada; Charlesworth (1994) at Auschwitz; Dewar and Fredericksen (2003) and Maxwell-Stewart (2013) in Australia; Arber (2013) in Norwich, UK; Ryan and Chard (2013) in Lancashire, UK; Smith and Buckley (2007) in New Caledonia; and Cooke (2000) and Zuelow (2004) at Kilmainham Jail, Dublin. In each of these places, the varying political, social and cultural contexts in which sites are embedded profoundly influence the ways in which they are treated in terms of the presentation and interpretation of their pasts, and the perspective taken on the current significance of the site. For example, Shackley (2001) argued that whilst Alcatraz's 'presentation' to visitors is heavily influenced by a Hollywood-inflected leisure and entertainment agenda, Robben Island is perceived to present penal (in) justice in a more 'serious' way. But just as questions are raised about the social construction of 'authenticity' at heritage sites of all kinds (e.g. DeLyser 1999) the marketing of prison sites as heritage places raises additional concerns about the ways in which some sites are perceived to portray 'authentic' or 'real' histories of punishment, whereas others are viewed as having a tendency to romanticise and distort, presenting an 'inauthentic' reading of carceral pasts.

Of particular concern in relation to the preservation and marketing of former prisons are the ways in which these processes are perceived to drive a wedge between the former function of the prison and contemporary discourses of imprisonment in the prevailing carceral system. Prison heritage sites tend to focus on presenting past prison practices, highlighting these for paying visitors as macabre, exotic, cruel, or harrowing, in ways which can, if they articulate any connection at all with the present, suggest that contemporary practices of imprisonment are somehow more 'civilised' or more 'lenient' than those of the past. However detailed their depiction of past punishment, prison heritage sites seldom attempt to situate themselves explicitly in relation to contemporary politics or practices of incarceration, and as Bruggeman (2012) noted in his study of the preservation of Eastern State Penitentiary in Philadelphia, US, prison heritage sites and museums face considerable challenges in articulating prison preservation with current carceral practices in any meaningful way.

Located in a city with one of the US' highest contemporary incarceration rates, just a few blocks from the type of ethnically diverse and economically marginalised neighbourhoods from which the majority of US prisoners are drawn, Bruggeman (2012: 174) described Eastern State Penitentiary as 'perhaps the most perfectly positioned [of all prison museums] to help its community make sense of the astonishing impact that mass incarceration has had on urban spaces in recent decades'. However, he concluded that for various reasons, it had largely failed to do so.

Writing in relation to the preservation of cultural landscapes, Hoelscher and Alderman (2004: 350) argued that elites 'often make or preserve … landscapes as a way to bolster a particular political order, and as a means to capital accumulation', and Bruggeman's (2012) analysis of the processes and circumstances behind the conservation of Eastern State Penitentiary suggests that the action of local elites in this case also (perhaps unwittingly) performed this function. Observing that 'museums are branded by the moment in which they are born', and that they 'carry with them the politics of those who shape their public roles' (ibid. 174), he noted that the committee of volunteers which formed to preserve the Penitentiary reflected the 'young, highly educated and predominantly white culture activists' (ibid. 177) who inhabited the newly gentrified local neighbourhood around the prison, and who had few connections either to prisoners who had been incarcerated at Eastern State, or to the criminal justice system in general. Although attempts were made to broaden the membership of the committee, funding problems meant that resourcing the pressing costs of preserving the site (even in a state of 'arrested decay') had to take precedence over fine-grained interpretation and community advocacy. The campaign to save the fabric of the site was successful, largely through the construction of a viable 'dark heritage' profile for Eastern State around 'jail break' parties, night-time tours and Halloween celebrations, but this was at the expense of 'calling the law into question' (ibid. 183). Thompson (2010) has argued that mass incarceration in the United States has disproportionately imprisoned young African American men, systematically criminalising urban space, and fuelling the contemporary race crisis, and Bruggeman (2012: 183) situated his study of Eastern State Penitentiary in this context, arguing that the 'moment' in which the prison museum was born was 'made possible by the racial politics of the US mass incarceration system, expressed in part through the reconfiguration of urban space during the late twentieth-century'. Eastern State offers a perception of 'authenticity' born of its macabre but saleable state of decay which reflects the interpretation of the site by those who preserved it, rather than one which draws on the living memories of local residents who were imprisoned there. The repurposing of prisons into prison heritage sites seems to perform a curious reversal in terms of the demographics of those who enter these spaces. As Bruggeman (2012: 172) noted, whereas 'American prisons are brimming with young black men, most in their late twenties', visitors to Eastern State, like museum visitors in general, tended to be 'well educated, age 50 or older, and almost always white'.

Very recently, *The Big Graph* installation (2014) at Eastern State Penitentiary, which depicted the US rising incarceration rate, and was paid for out of funds raised through prison tours, is perhaps an indication that Eastern State is increasingly using its position to articulate a message about contemporary confinement. 'Not all our visitors want to have these conversations, and that is completely fine with us', Eastern State Penitentiary's vice president, commented to the media. 'We'll still tell the stories of Al Capone's incarceration and the stories of escapes. But we also are thinking this is the place for people who want to see the huge changes in the US criminal justice system. This is the place to have those conversations'.[1]

The commodification of the macabre at prison sites erases as much as it reveals in relation to the communication of the meaning and purpose of contemporary imprisonment and punishment to visiting audiences (Morin 2013, Walby and Piché 2011). Where prison heritage sites *do* articulate a consciously and intentionally politicised message through the preservation and interpretation of the site, it is rarely about imprisonment *per se*. The 'Stasi Prison', for example, at Hohenschönhausen, in former East Berlin, was a place of incarceration for East German political prisoners, closed after German reunification and now preserved as a heritage site (see Figure 11.6). In on-site interpretation it is narrated as a means of understanding the operation of the Stasi[2] and the nature of state socialism, as part of a wider process of the rehabilitation and acknowledgement of twentieth-century German history (see Figure 11.7). Since imprisonment itself, at this site, was politically motivated under a now defunct political system, there is no connection made to contemporary imprisonment in Germany. Similarly, at Kilmainham Gaol in Dublin, despite the long history of the site as an 'ordinary prison', the dominant narrative presented through the restoration and preservation of the site is of its role as a place of incarceration of the leaders of the 'Easter Rising' of 1916 against British rule, and of the subsequent executions which contributed to the crystallisation of Irish nationalist resistance and Republicanism. Closed by the Irish Free State in 1924, Kilmainham initially fell into disrepair, but is now a major tourist site, discursively linked directly to narratives of Irish history and national identity, rather than to discourses of contemporary imprisonment in Ireland.

---

1   Newsworks, 4/6/2014, http://www.newsworks.org/index.php/homepage-feature/it em/68793-at-eastern-state-massive-sculpture-illustrates-exploding-us-incarceration-ra te?linktype=hp_impact, accessed 16/6/2014.

2   The German Democratic Republic's Ministry for State Security (*Ministerium für Staatssicherheit*), was commonly known as the *Stasi* (*Staatssicherheit*). It was the state security service, or 'secret police' of the GDR.

Figure 11.6　The 'Stasi Prison', at Hohenschönhausen, former East Berlin

**Figure 11.7　Preserved 'interview room' at the 'Stasi Prison',
Hohenschönhausen, former East Berlin**

Although prison museums offer opportunities to engage visitors in discourses around contemporary incarceration, these opportunities tend to be lost either because of an overriding focus on a specific prisoner, event, or context for which a site has become known (as at Robben Island, Kilmainham, or at Hohenschönhausen), or because of an explicit or implicit portrayal of penal reform, which constructs the present as 'enlightened' and 'civilised' when compared to a barbarous past, and which simultaneously creates a social distance between 'us' and 'them', the visitor and the prisoner, in ways which reinforce an erroneous impression that imprisonment is somehow disconnected from society at large (as previously at Eastern State Penitentiary).

In the context of a postmodern 'New Museology' which advocates open texts which refuse resolution and promote ambiguity and textual possibilities for the reader/visitor, Garton-Smith (2000: 12) argued that prison museums' lack of complex interpretation is puzzling. The 'ideal' text is non-authoritarian, raises questions rather than answering them and places the visitor at the centre of the meaning process, and she argued that prison museums should be productive of rich texts which enable visitors to think about difficult issues. Observing the disappointing 'failure to use former prison sites as places for the articulation of current penal issues' she argued that by presenting past forms of imprisonment, prison heritage spaces are limited in the potential complexity of their exhibition of the site's history, in therefore in the ability of the site to comment on contemporary issues (2000: 7 and 11). The 'freeze' she described as a result of foregrounding the physical and historical fabric of the site 'seems to make logical the separation of the exhibition from wider issues'.

In her work on Australian prison museums, Wilson (2008) similarly traced the contexts in which former prisoners' voices are silenced in the narration and interpretation of prison sites. She noted that the prominence of prison officers within stakeholder groups who take on significant roles in the preservation of sites and exert considerable influence over their presentation, tends to deliver to visitors a negative perception of the former inmates of the prison. Likewise, a focus in interpretation on specific 'celebrity' prisoners tends, she argued, to create a 'broadly tolerant, perhaps even approving sensibility' (ibid. 2008: 218), on the part of visitors towards these 'celebrities', and to exclude still further the stories of 'ordinary' criminals from the articulation of the prison's history. As Flynn (2011) also observed in relation to the negotiations over the future of Long Kesh/The Maze, local perspectives from community-level and other stakeholders, such as former prisoners, are often overlooked.

Prison Towns

Prisons may be the quintessential carceral landscapes, but just as prisons are porous in terms of the movement inside and out of inmates, staff, goods and services, communications and so on, they are also porous in that their carceral nature pervades the locale, particularly in places which have come to be defined by imprisonment – prison towns. The phenomenon of prison towns has recently been observed and studied in the United States, where the prison boom has brought correctional facilities to innumerable settlements afflicted by economic disadvantage, many to towns which already had one or more prison in their hinterland. The effect of this kind of prison clustering in the US, and the inscription of incarceration on the wider cultural landscape, are only recently starting to be reported and researched, but elsewhere, prison towns, or penal zones, have already received academic attention.

In the Russian Federation, a penal sub-region of the Republic of Mordovia is one of several areas which have specialised in imprisonment as an economic function since the 1930s inception of the Soviet Gulag (Pallot et al. 2010). With a free population of approximately 18,000 living in 10 neighbouring settlements, the remote, forested sub-region is known locally as the 'zone' (*zona*) or 'regime zone' (*rezhimnaya zona*). 15,000 prisoners are housed in several different types of institutions, strung out along an arterial road which links the settlements. As Pallot et al. (2010: 14) noted, in the minds of family members making prison visits, the area recalls Solzhenitsyn's *Gulag Archipelago*. Amongst the prison colonies, with their distinctive high fences, watch towers, barbed wire and locked gates, the settlements inside the 'zone' contrast starkly with those outside of it. The prison system has brought well developed infrastructure, out of reach of the marginalised rural settlements beyond the 'zone'; mobile telephone signal, mains gas and piped water, kindergartens, schools, a sports centre, and a modern hospital. Within the 'zone' there is little employment that does not originate in the prison system in one form or another, and many of the prison personnel are descendants of Soviet and Gulag era prison guards whose families settled in the region decades ago. In this curiously isolated and introspective community, the 75th anniversary of the founding of the penal region under Stalin was celebrated in 2006 by a parade, a dance in the local social club, refurbishment of the local museum of the prison system, and the publication of commemorative poetry in the institutional newspapers. As well as enjoying a higher quality of life than is possible locally, prison employees in the 'zone' are also encouraged to feel an attachment to, and a pride in, the penal region, culminating in a sense of loyalty to the 'Little Motherland' (ibid. 23), supported by a historical narrative which valorises service, patriotism and duty, and which remains largely silent on the alternative histories of Soviet and Russian imprisonment from the perspective of the incarcerated.

Whereas the specific identity of the penal sub-region in Mordovia has become sedimented over decades and generations, in the United States, the phenomenon of prison towns with clusters of correctional facilities is very much more recent. However, in different ways and with different cultural connotations, incarceration is equally culturally significant. Although they exist in starkly different contexts, Fraser's (2000) vivid description of the prison town of Susanville, in the high desert of northeast California, bears some comparison with Mordovia. Prisons are relative newcomers to Susanville, the first in 1963, and then resolute expansion in the 1990s, but with them has come a similar development of infrastructure, higher living standards, and an increasingly familiar, everyday presence of prison personnel in local shops, bars and fuel stations; 'the overwhelming presence of men with military haircuts and trim moustaches, the constant talk of prison scandals and violence' (Fraser 2000: 777). Although prison employees in Susanville seem to 'blow through the town' (ibid. 783) on their paths to career advancement, rather than becoming rooted in the area as in the case of Mordovia, perhaps this is a function of time.

Fraser (2000) described Susanville from the perspective of an 'outsider' to the prison system, someone who grew up there before the prisons came in earnest and who recognised the difference they had made to the town. Her reflections emphasised the prominent place of the prisons in the local landscape:

> High Desert and CCC [California Correctional Center] come into view on the right. Together, the prisons occupy nearly 1000 acres of desert, set on a terrain that rolls and spreads like an unnamed planet for the mad and the lonely. A flower here seems like an alien and wondrous thing. The institutions not only confirm but seem to incorporate the surrounding desolation, blending into the distant desert hills ... (ibid. 791)

> It is ironic that the most concrete example of the change in Susanville – the prison itself – takes on its own abstract symbolism. During the day, no one could mistake the prison for anything but what it is, with its gray cement structures, high fencing with spiralling razor wire, guard towers with tinted glass – and silence.

> But it is at night that the prison seems to take on a life of its own. At night, the proximity of the town to the prison is most evident because of the lights. Since more than half of the inmates are Level III and IV, the prison is surrounded by thirty-foot high poles topped with glaring amber lights. The resulting glow changes the night sky, affects the entire county – the yellow can be seen from fifty miles away. . On overcast nights, the clouds reflect the prison's light, casting the sky into an eerie, hellish spectacle, like embers from a tremendous firestorm, fallout from a war. (ibid. 794)

For Fraser (2000), the lights came to represent the 'ultimate, and intangible, cost, of High State Desert Prison' (ibid. 795), with another resident describing the 'prison, bright as day' as 'futuristic, unnatural, something out of a science fiction movie. Like some giant alien mother ship had landed' (ibid. 795).

Fraser's account of the coming of the prisons to Susanville is intended to be a cautionary tale of the impact of this industry on the landscape and the atmosphere of small rural towns like hers, and recent scholarship would suggest that the experience of Susanville; 'not just a small town but The American Small Town' (2000: 778) mirrors that of innumerable others, including Cañon City,

Colorado, which a sign proclaims 'Corrections Capital of the World', and whose mayor reportedly boasted of the, 'nice, non-polluting, recession-proof industry here' (Brooke 1997: 20; in Welch and Turner 2007: 57). The most populous city in Fremont County Colorado, and site of the Colorado Territorial Correctional Facility, the name of Cañon City has become synonymous with incarceration. Nearby Florence is the site for the Federal Correctional Complex for male inmates, operated by the US Federal Bureau of Prisons, consisting of a medium-security, a high-security, and the only federal Supermax facility, and elsewhere in Fremont County there are further facilities including Fremont Correctional Facility, Arrowhead Correctional Center, Skyline Correctional Center, Four Mile Correctional Center, and Centennial Correctional Facility, as well as numerous facilities for migrant detainees (Doty and Wheatley 2013). In a familiar story of economically depressed rural towns, the area campaigned for the recruitment of the prisons, which were expected to bring work and development (Perkinson 1994).

Cañon City's notoriety as a prison town is now underscored by an award-winning online documentary 'Prison Valley Colorado'[3] which describes it as 'a town in the middle of nowhere, with 36,000 souls and 13 prisons, one of which is Supermax, the new "Alcatraz" of America … A journey into what the future might hold'. The online tour of 'Prison Valley Colorado' opens with views of the approach to the area, with a voiceover describing Cañon City as 'some godforsaken place', 'a clean version of Hell' with an economy so reliant on prisons that 'even those on the outside, are inside'. The website is an example of non-linear multimedia, with immersive interactivity, where the viewer controls their own journey, learning new information along the way. Navigating within the website, visitors are taken through a motel, introduced to prison reformers, local shopkeepers, and correctional officers, within a site structure that allows them to guide their own 'tour' of 'Prison Valley' and its inhabitants, with movies and still life photos of urban landscapes and prisonscapes (with no prisoners' faces), voiceovers, interviews, discussion fora and so on. The central premise of the documentary, according to the director David Dufresne,[4] was to 'show a system without judging anyone', and to represent the prison industry without lapsing into voyeurism, but he acknowledged that doing so was a challenge, given the 'deeply aesthetic' world of the prison, with 'its straight lines, blocks, bars and, in this case, the striped uniforms and corridors we know so well from the movies'. Whilst the online 'tour' has its own appeal, it must be viewed in light of the recent critical discourse over 'carceral tours' (Piché and Walby 2009, 2010 and 2012, Wilson et al. 2011, Dey 2009, Smith 2013) which calls into question the validity

---

3   A production with photography by Philippe Brault in collaboration with author and co-director David Dufresne and produced by Alexandre Brachet and Gregory Trowbridge for Arte. Prison Valley is an interactive production that explores the prison industry in Cañon City, Colorado in the US, http://prisonvalley.arte.tv/?lang=en, accessed 22/1/2014.

4   http://prisonvalley.arte.tv/blog/en/2010/02/19/%c2%abwhen-lots-of-different-experiences-suddenly-come-together%c2%bb/, accessed 24/1/2014.

of observations of places of confinement made during what are argued to be highly controlled, scripted and regulated visits, and as such to afford little insight into the nature of imprisonment.

Thus far this chapter has explored carceral landscapes in terms of physical spaces of incarceration; the ways in which they serve as physical manifestations of discourses of justice and retribution, punishment and rehabilitation, and how, whether as prisons or post-prisons, they reflexively shape the communities which interact with them, as social, cultural and political elites promote their own values through the inscription of carceral sites. Whereas 'Prison Valley Colorado' takes inspiration from and is anchored in a really-existing carceral landscape, for the remainder of the chapter the focus shifts to consider virtual carceral landscapes which exist nowhere except in the minds of those who create them, and via the software they use to give form to their visions.

'Virtual' Prisons and the 'Spectacle' of Punishment

Landscapes are not always physical and tangible; they can also be virtual and imagined. Like the self-directed virtual tour of Cañon City via 'Prison Valley Colorado', some virtual carceral landscapes draw viewers into vicarious constructions of carceral landscapes which demand reflection on ideas of punitivity connected to public opinion about imprisonment (discussed in Chapter 10).

One such example is a computer game about building prisons. The 'Gamer's Hub' recently previewed UK Introversion Software's *Prison Architect*, on display at the Eurogamer Expo at Earl's Court, London;[5] 'In it, you're handed a prison warden's truncheon and the responsibility for managing the day-to-day to-and-fro of the goings on within your jail ... the aim is to build an economically-viable business, while meeting the needs of inmates and investors alike'. Gamers create a prison in their own image, giving the institution the facilities it needs, 'from cells and generators to toilets and adequate lighting' with the opportunity to construct 'an execution chamber for a waiting inmate, guilty of the murder of his wife and her lover'. Whilst the gamer designs the space, 'he and a priest sit in one of the cells awaiting the inevitable. As you complete each rudimentary objective, brief flashbacks of his path to the pen are recalled – polaroid snapshots and comic-book stills capture the moments before his arrest, as the prisoner tells of his motives, malice and regret'.

One of the designers behind the game admitted in a 2012 interview that Introversion had not fully considered the contentious nature of prisons, especially in the US:

> I think they have a very different view on incarceration than we do in the UK ...
> We're not trying to stamp down on our own views of prisons and incarceration,

---

5   http://thegamershub.net/2012/10/prison-architect-preview/, accessed 22/1/2014.

but we want to make an accurate-ish model where you can explore punishment vs. rehabilitation, those sorts of things. Learning quite quickly that we didn't have an understanding of all this, we reached out to quite a prevalent rehabilitated prisoner and currently serving prison officers to talk to them about whether there was anything ridiculous in our game. We're not trying to make a serious model for the Home Office. It's a game. But it's also an interesting and in-depth project.[6]

Human geographers have recently begun to explore virtual landscapes such as Second Life, with for example Li et al. (2010) discussing the notion of the 'multiple spaces' in which we live, some of which are virtual social worlds far beyond computer games. They examine the interplays and connections among these different spaces, and their social implications. In terms of *Prison Architect*, although the potentially controversial nature of the game's subject matter appears to have escaped the attention of its designers until rather late in the day, perhaps there is more to this than meets the eye. Although in its early stages of development and release *Prison Architect* offers few variations on a predictable theme of prison design, apparently as it develops further there will be more 'political' choices to make; the Games Hub reviewer was told by the designers that 'we can expect much more licence to build a slammer in our own moral image further down the line ... we can expect anything from Darth Vader style dungeons to left-wing, liberal holiday homes – whichever best suits your mood'. However, by maintaining the economic viability of the prison as the overall logic of the game, *Prison Architect* raises interesting questions about the view of prisons and imprisonment held by the general public, and the extent to which the game panders to 'presumably punitive' public opinion.

Fiddler (2012: 2) pointed out that many contemporary media challenge the messages projected by 'standard' representations of imprisonment such as print media, forcing us to 'look anew'. For carceral geographers interested in the construction and experience of carceral cultural landscapes, and understandings of them outside of the context of imprisonment, *Prison Architect* is a not just a representation of prison life created as spectacle for the entertainment of an audience, with the potential to shape the views and opinions that they hold: it requires the active and interested participation of the audience in designing the penal space itself; arguably the experience is reflexive, enabling experimentation and reflection. In any case, this game offers the opportunity to consider virtual carceral space as a one of the 'multiple spaces' in which we live, the interplay and connections between this and other lived spaces, and the social implications of that interplay, and in so doing, makes a direct link to the wider discourses of incarceration.

---

6   http://www.pcgamer.com/2012/09/26/prison-architect-paid-alpha-released-for-good-behaviour-trailer-and-interview-within/, accessed 22/1/2014.

Conclusion: Beyond a 'Gaze' on the Spectacle of Punishment

Foucault's (1979) familiar contention is that the prison replaced the public spectacle of punishment, as the gallows, the stocks, and public humiliation through punishment wrought against the body was replaced by internalisation of the carceral regime; through regulation of space, segregation of individuals, and unseen but constant surveillance of the body, the subject is moulded into its own primary disciplinary force. An equally familiar argument within contemporary criminology and media studies is that the 'spectacle' of punishment is returning in new guises, through media representations of incarceration which make the closed world of the prison open to public view, albeit in selective and incomplete ways, through documentaries, docu-dramas, movies, novels, and so on (e.g. Turner 2013b, Mason 2013). In a special issue of the *Prison Service Journal* on representations of imprisonment, Kearon (2012) examined the ways in which fictional accounts of imprisonment intersect with dominant narratives within news media. Ranging from the fictional to the 'real-life', all necessarily present constructions, interpretations and partial readings of their subject matter. But all are acknowledged to be important in shaping public opinion of imprisonment, which arguably in turn shapes punishment policy, in contexts in which criminal justice policy is politicised to the extent that criminal justice becomes a political tool rather than a balanced assessment of the effectiveness of interventions (Cheliotis 2010). As Lappi-Seppälä (2002: 33) observed, in these contexts (such as the United States and UK), 'the higher the level of political authority, the more simplistic the approaches advocated. The results can be seen in slogans that are compressed into two or three words, including "prison works", "war on drugs" and "zero tolerance"' which in turn leads politicians to 'pander to punitive (or presumably punitive) public opinion with harsh tough-on-crime campaigns'. Where prison policy is informed less by an understanding of the likely success of specific interventions for the stated aims of incarceration, than by a political imperative to respond to public opinion, or as Gilmore (2002: 16) has argued, to use the prison system as 'a project of state-building', media representations of imprisonment take on a real significance.

A spectacle, of course, requires a gaze, and visitors to prison sites, be they online virtual landscapes, prison museums, repurposed sites or macabre decaying prison buildings, have much in common with the tourists whose 'gaze' is the focus of geographical research examining their ways of seeing, and the power inherent both in their gaze and in the manipulation of representations and experiences (Urry 1990, Gibson 2010). Academic approaches to tourism have recently diversified away from the primacy of the 'gaze' to encompass other senses, and to analyse tourism encounters as affective and embodied, in parallel with the recent reinvigoration of landscape studies within human geography, itself connected with the incorporation of notions of materiality, embodied practice and perception, and the interweaving of self and landscape through a focus on their performance via everyday embodied and imaginative practices.

Visits to prison sites may elicit emotional responses and bring to mind questions of morality, and embodied, tactile and material encounters may heighten what Waitt et al. (2007: 261) have described as the 'drama of encounter'. Such considerations underpin recent developments in work on the interpretation and presentation of prison sites, which has started to go beyond critiques of the ways in which they interpret history, and to consider more directly the ways in which such sites 'provide a set of narratives through which to understand the present meanings of imprisonment and punishment' (Walby and Piché 2011: 453). Commenting on the partiality and concealment of tourism encounters, Robinson (2001: 54) noted that few tourists pursue total immersion in a different culture, instead seeking 'safe glimpses of cultural difference' and often being 'satisfied with simulacra'; and this characterisation *appears* equally applicable to encounters with prison landscapes. Although the *presentation* of numerous prison sites has been scrutinised, as Walby and Piché (2011: 451) observed, they create 'a polysemy of meanings ... with critical, indifferent and punitive interpretations ... possible'.

Like other 'dark' sites, prisons are open to interpretation; however, very little research has engaged with this process of interpretation on the part of visitors themselves. These interpretations bear further exploration, and as Turner (2013a: 41) has argued, cultural geography is well-placed to theorise the affective nature of imprisonment, and the embodiment and performance of relationships between prison and society.

Chapter 12

Afterword

Carceral geography may at first appear rather 'niche'. Although prisons and criminal justice systems are integral parts of governance and techniques of governmentality, the geographical study of the prison and other confined or closed spaces is still relatively novel. That said, carceral geography has already made substantial progress, has already established useful and fruitful dialogues with cognate disciplines of criminology and prison sociology, and is attuned to issues of contemporary import such as hyperincarceration and the advance of the punitive state. This volume has sought to provide an overview of that progress and of the scholarship which has so far defined carceral geography, in light of the areas of convergence with work in other disciplines.

Although geographical research into the spaces and practices of incarceration has often taken a fine-grained, micro-level approach to carceral spaces, looking at individual experiences, and using methodologies which resonate with prison ethnographies (Wacquant 2002, Drake and Earle 2013), these studies have relevance far beyond the individuals and institutions within their purview. Rather than being a 'niche' pursuit, carceral geography has the opportunity to use the carceral context as a lens through which to view concepts with wider currency within contemporary and critical human geography. Thus far, as the foregoing chapters in this book outline, carceral geography has made key contributions to debates within human geography over mobility, liminality, and embodiment, and it has increasingly found a wider audience, with geographical approaches to carceral space being taken up by and developed further within criminology and prison sociology (e.g. Crewe et al. 2014).

What it is that carceral geography brings to the study of prisons and imprisonment is geography's understanding of space, an understanding developed over a century of debate over the nature of geographical space. Essentially, rather than seeing space as *absolute*, in that it exists independent of other objects and/or relations, and acts as a container for things and processes, contemporary human geography views space as *relative*, and increasingly as *relational*. Whereas *relative* space can be defined *only* in relation to objects and processes, and involves the explanation and representation of the relationships between spaces and objects and processes, *relational* space dissolves the boundaries between objects and space completely. It invokes an understanding of objects and processes *as* space, and of space *as* objects and processes, each understandable only in relation to the other in a perpetual process of becoming. The advantage of this approach to space is that, rather than treating space as something which can be classified, delimited and pigeon-holed, 'thinking space relationally' presents a challenge to geographers to consider space as 'encountered, performed and fluid' (Jones 2009: 492). Spaces are seen as 'open, discontinuous, relational and internally diverse' (Allen et al. 1998: 143), and as 'complex and unbounded lattice[s] of articulations' (ibid. 65).

What this has essentially meant is that rather than seeing prisons as spatially fixed and bounded containers for people and imprisonment practices, rolled out across Cartesian space through prison systems straightforwardly mappable in scale and distance, carceral geography has tended towards an interpretation of prisons as fluid, geographically-anchored sites of connections and relations, both connected to each other and articulated with wider social processes through and via mobile and embodied practices. Hence the focus on experience, performance and mutability of prison space, the porous prison boundary, mobility within and between institutions, and the ways in which meanings and significations are manifest within fluid and ever-becoming carceral landscapes.

There is further potential for confluence, though; for example, carceral geography has contributions to make to the emergent discourse of criminological cartography, in which the 'critical cartography' long familiar within human geography (e.g. Harley 1988, Crampton 2009) is now being absorbed by criminologists utilising maps to generate empirical insights and promote social justice. Although traditionally applied to crime mapping, the communicative power of cartography is already being recognised through projects such as '*Prison Map*'[1] which uses satellite imagery to depict the US carceral estate, making visible 'this hidden penal architecture, and with striking effect; a surreal and seemingly endless grid of satellite photographs, a dystopian archipelago of incarceration' (Kindynis 2014: 236). Recognising the invaluable contribution of radical mapping projects that work against political processes of erasure and disconnection, Gill et al. (forthcoming) call for a broadened and strengthened role for what they call 'dynamic critical peno-cartography'. Kindysis (2014) urges criminologists to critically re-engage with cartography as an object of research, and carceral geography is well placed to contribute to this engagement.

Beyond the existing dialogues with criminology and prison sociology, there is also considerable potential for further transdisciplinary synergies between carceral geography and psychology, and architectural studies, in relation to prison design and the lived experience of carceral spaces. Whereas in previous research into the prison environment, environmental psychology has tended towards a limited range of largely quantitative, deterministic studies of 'response' to carceral environments (Schaeffer et al. 1988, Tartaro 2003, Morris and Worrall 2010, Houston et al. 1988), carceral geography has the potential to nuance these approaches significantly by considering the mediation of this response by human subjectivities, in approaches consonant with the growing body of work on affect and emotion in contemporary human geography.

There are also further areas of potential confluence between carceral geography and other subdisciplines of human geography, where there is considerable scope for

---

1   www.prisonmap.com, accessed 24/2/2014.

further research inquiry. Given the profound importance of time in carceral space, understandings of carceral TimeSpace could be pursued in greater depth, and with useful crossover into other disciplinary areas. There is also enormous scope for further work in relation to the embodied experience of incarceration, engaging with theorisations within feminist and corporeal geographies, and investigating further the operation of stigma in relation to carceral contexts. Although research into carceral systems has already contributed significantly to the awareness and theorisation of coerced, governmental or disciplined mobility, the immense potential for research into the practices and experiences of these forms of carceral mobility, remains largely untapped, and the movement of pre-trial detainees from police custody, through remand facilities, to trial and to correctional facilities, as well as the process of 'ghosting' inmates between correctional facilities, are yet to be fully explored. And although the nature of the prison boundary has already become established as a strong theme of carceral geography research, drawing on critical border studies could provide a new perspective on this issue, in relation to bordering practices and experiences. In relation to the punitive turn, there is clearly the potential for carceral geographers to further investigate phenomena such as the million-dollar blocks in the US, and the relationship between the carceral and the state in a range of locales within the extended reach of the prison system, in relation to urban and racialized marginality. In dialogue with architectural and cultural geographies, research could further scrutinise the buildings which constitute the carceral estate, as manifestations of penal philosophies, and as sites with meaning beyond their functionality as places of incarceration (Jewkes and Moran 2014).

Carceral geography has already established itself in relation to prison research in other disciplines, but there is considerable scope for it to contribute to the advancement of geographical scholarship, bringing to bear a perspective on apparently 'closed' spaces that has much to offer to critical human geography, both in relation to a new 'territory' for exploration, and to regime shifts in the carceral landscape. The challenge is for carceral geography to contribute to contemporary discourse within human geography, to remain open to transdisciplinary engagement, and to retain a critical perspective that helps bring about progressive social transformation.

That progressive social transformation is a compelling motivation for many carceral geographers, who are active beyond the academy in trying to bring about change in carceral systems, be the focus on mainstream imprisonment, or on migrant detention. This volume has taken as its remit the academic expression of the work of scholars of incarceration, but it is critical to observe that outwith their published work, many are also committed activists. Although the sphere of academic research and publishing engages issues of incarceration on one level, there is much more to be done, and scholars in this field often tread a fine line between maintaining research access to carceral institutions, and critiquing their activities. Underpinning much of the scholarship reviewed here is an abolitionist praxis, in which abolition means a situation in which prisons, policing and the carceral system are not deployed to address what are at their essence, social, political and economic problems. Or, as Lappi-Seppälä (2002: 33) succinctly put it, solutions are not 'sought where they cannot be found – the penal system'. It is imperative that prison scholars maintain a critical perspective, addressing not only the lived experience of incarceration, the geographies of carceral systems and the relationship between the carceral system and the state, but keeping a steady focus on the locus of power which shapes and reshapes the system itself, and which moulds punitive opinion in ways which serve to sustain incarceration. In practical terms, this means that carceral geographers must maintain a clear vision of their research integrity, as they navigate the complex webs of commitments and connections which inevitably develop within and beyond carceral spaces. All researchers owe a debt to those whose opinions and experiences they draw upon, but in the context of incarceration, they must carefully manage and maintain both academic engagement, and advocacy for the confined.

With this in mind, this volume has opened a space for a more attentive consideration within carceral geography of the *purposes* of imprisonment and the contexts in which these are socially constructed. In its various chapters it has called into question the intentions behind aspects of carceral spaces and practices, from the nature of prisoner accommodation and the ways in which it enables or prevents experience of privacy or solitude; to the location of prison facilities, the logics behind siting policies, and the implications of these decisions for the experience of prisoners, their families, and host communities; through to the design of prison spaces and the material manifestation of penal philosophies. By focusing on fluid and relational carceral spaces, be they physical, virtual or embodied, carceral geography has an opportunity to bring a fresh perspective to criminal justice practices as they pertain to punishment by imprisonment, just as incarceration provides a new and challenging context in which to think through theoretical advances in contemporary human geography.

The intention here is not to suggest that geography is a hidden explanatory lens through which prisons can somehow be 'understood', but rather that space as understood by geographers is a means, at times evident only in the margins or in the background of research in criminology and prison sociology, that could usefully be afforded more prominence. Carceral geography has much to learn from longer-standing engagements with incarceration, but they too have much to learn from geography.

Bibliography

Abrams, K. S. and W. Lyons 1987. *Impact of correctional facilities on land values and public safety.* North Miami, FL, FAU-FIV Joint Center for Environmental and Urban Problems.

Aday, R. H. 1994. Aging in prison: A case study of new elderly offenders. *International Journal of Offender Therapy and Comparative Criminology* 38(1): 79–91.

Adey, P. 2006. If Mobility is Everything Then It Is Nothing: Towards a Relational Politics of (Im)mobilities. *Mobilities* 1(1): 75–94.

Adey, P. 2008. Airports, mobility and the affective architecture of affective control. *Geoforum* 39: 438–451.

Adey, P., L. Brayer, D. Masson, et al. 2013. 'Pour votre tranquillité': Ambiance, atmosphere and surveillance. *Geoforum* 49: 299–309.

Aebi, M. F. and Kuhn, A. 2000. Influences on the prisoner rate: Number of entries into prison, length of sentences and crime rate. *European Journal on Criminal Policy and Research* 8(1): 65–75.

Agamben, G. 1998. *Homo Sacer: Sovereign Power and Bare Life.* Stanford University Press, Stanford.

Agamben, G. 2005. *State of Exception.* University of Chicago Press, Chicago.

Ahmed, S. and J. Stacey (eds) 2001. *Thinking through the Skin.* Routledge, London.

Alexander, M. 2010. *The New Jim Crow: Mass Incarceration in the Age of Colorblindness.* The New Press, New York.

Alford, C. F. 2000. What would it matter if everything Foucault said about prison were wrong? *Discipline and Punish* after twenty years. *Theory and Society* 29(1): 125–146.

Allen, J. 2004. The Whereabouts of Power: Politics, Government and Space. *Geografiska Annaler B* 86(1): 19–32.

Allen, J. 2006. Ambient power: Berlin's Potsdamer Platz and the seductive logic of public spaces. *Urban Studies* 43: 441–455.

Allen, J., D. Massey and A. Cochrane 1998. *Re-thinking the Region.* Routledge, London.

Allspach, A. 2010. Landscapes of (neo)liberal control: The transcarceral spaces of federally sentenced women in Canada. *Gender, Place & Culture* 17(6): 705–723.

Altheide, D. L. 1991. The Mass Media as a Total Institution. *Communications* 16(1): 63–72.

Anderson, B. 2006. Being and becoming hopeful: Towards a theory of affect. *Environment and Planning D: Society and Space* 24: 733–752.

Anderson, K. and S. Smith. 2001. Emotional geographies. *Transactions of the Institute of British Geographers* 26: 7–10.

Arber, N. 2013. Presenting Prison History at Norwich Castle. *Prison Service Journal* 210: 29–33.

Aretxaga, B. 1995. Dirty Protest: Symbolic Overdetermination and Gender in Northern Ireland Ethnic Violence. *Ethos* 23: 123–148.

Armstrong, S. 2012. 'Siting Prisons, Sighting Communities: Geographies of Objection in a Planning Process', http://ssrn.com/abstract=2117840.

Baer, L. D. 2005. Visual Imprints on the Prison Landscape: A Study on the Decorations in Prison Cells. *Tijdschrift voor Economische en Sociale Geographie* 96(2): 209–217.

Baer, L. D. and B. Ravneberg 2008. The outside and inside in Norwegian and English Prisons. *Geografisker Annaler B* 90(2): 205–216.

Ball, K. S. 2010. Workplace Surveillance: An Overview. *Labor History* 51(1): 87–106.

Barnett, C. 2005. The consolations of 'neoliberalism'. *Geoforum* 36(1): 7–12.

Bauman, Z. 2000. Social Uses of Law and Order. In *Criminology and Social Theory*, edited by D. Garland and R. Sparks. Oxford University Press, Oxford.

Beck, U. 1992. *Risk Society.* Sage, London.

Becker, H. S. 1964. Personal change in adult life. *Sociometry* 27(1): 40–53.

Beckett, K. and N. Murakawa 2012. Mapping the shadow carceral state: Toward an institutionally capacious approach to punishment. *Theoretical Criminology* 16(2): 221–244.

Beckett, K. and S. Herbert 2010. Penal Boundaries: Banishment and the Expansion of Punishment. *Law & Social Inquiry* 35(1): 1–38.

Bedard, K. and E. Helland 2004. The location of women's prisons and the deterrence effect of 'harder' time. *International Review of Law and Economics* 24: 147–167.

Benoliel, J. Q. 1990. Commentary: Effects of a Double-Bind Environment. *Rehabilitation Nursing* 15(4): 200–201.

Benyamin, Y., H. Leventhal and E. A. Leventhal 2004. Self-rated oral health as an independent indicator of self-rated general health, self-esteem and life satisfaction. *Social Science & Medicine* 59: 1109–1116.

Bergson, H. 1911. *Matter and Memory.* Swan Sonnenschein, London.

Bevir, M. 1999. Foucault and critique; deploying agency against autonomy. *Political Theory* 27(1): 65–84.

Biggam, F. H. and K. G. Power 1997. Social support and psychological distress in a group of incarcerated young offenders. *International Journal of Offender Therapy and Comparative Criminology* 41(3): 213–230.

Bigo, D. 2007. Exception et ban: A propos de l'Etat d'exception. *Erytheis* 2: 115–145.

Binswanger, L. 1984. Dream and existence. *Review of Existential Psychology & Psychiatry* 19(1): 81–105.

Blankenship, S. E. and E. J. Yanarella 2004. Prison recruitment as a policy tool of local economic development: A critical evaluation. *Critical Justice Review* 7(2): 183–198.

Blomley, N. 2005. Flowers in the bathtub: Boundary crossings at the public-private divide. *Geoforum* 36: 281–296.

Bondi, L. 2005. Making connections and thinking through emotions: Between geography and psychotherapy. *Transactions of the Institute of British Geographers* 30(4): 433–448.

Bonds, A. 2006. Profit from Punishment? The politics of prisons, poverty and neoliberal restructuring in the rural American Northwest. *Antipode* 38(1): 174–177.

Bonds, A. 2009. Discipline and Devolution: Constructions of Poverty, Race and Criminality in the Politics of Rural Prison Development. *Antipode* 41(3): 416–438.

Bonds, A. 2012. Building Prisons, Building Poverty: Prison Sitings, Dispossession, and Mass Incarceration. In *Beyond Walls and Cages: Prisons, Borders, and Global Crisis*, edited by J. Loyd, M. Mitchelson and A. Burridge. University of Georgia Press, Athens, pp. 129–142.

Bonds, A. 2013. Economic Development, Racialization, and Privilege: Yes in My Backyard Prison Politics and the Reinvention of Madras, Oregon. *Annals of the Association of American Geographers* 103(6): 1389–1405.

Bourdieu, P. and L. J. Wacquant 1992. *Réponses: Pour une anthropologie réflexive.* Seuil, Paris.

Boyer, E. E. and N. J. Nielsen-Thompson 2002. A comparison of dental caries and tooth loss for Iowa prisoners with other prison populations and dentate U.S. adults. *Journal of Dental Hygiene* 76(2): 141–150.

Braudel, F. 1980. *On History.* Chicago University Press, Chicago.

Brodie, A., J. Croom and J. O. Davies 1999. *Behind Bars: The Hidden Architecture of England's Prisons.* English Heritage.

Brodie, A., J. Croom and J. O. Davies 2002. *English Prisons: An Architectural History.* English Heritage.

Brooke, J. 1997. Prisons: A Growth Industry. *New York Times* (November 2): 20.

Brown, A. and A. Barton 2013. The Prison and the Public: Editorial Comment. *Prison Service Journal* 210: 2–4.

Brown, M. and Marusek, S. 2014. 'Ohana Ho 'opakele: The Politics of Place in Corrective Environments. *International Journal for the Semiotics of Law-Revue internationale de Sémiotique juridique* 27: 225–242.

Bruggeman, S. 2012. Reforming the Carceral Past: Eastern State Penitentiary and the Challenge of the Twenty-First-Century Prison Museum. *Public History* 113: 171–186.

Brunovskis, A. and R. Surtees 2008. Agency or Illness – The Conceptualization of Trafficking: Victims' Choices and Behaviors in the Assistance System. *Gender Technology and Development* 12(1): 53–76.

Burayidi, M. A. and M. Coulibaly 2009. Image Busters: How Prison Location Distorts the Profiles of Rural Host Communities and What Can be Done About It. *Economic Development Quarterly* 23: 141–149.

Burnett, R. and S. Maruna 2004. So 'prison works', does it? The criminal careers of 130 men released from prison under Home Secretary, Michael Howard. *The Howard Journal of Criminal Justice* 43(4): 390–404.

Butcher, S. 2012. Embodied cognitive geographies. *Progress in Human Geography* 36(1): 90–110.

Cadman, L. 2010. How (not) to be governed: Foucault, critique, and the political. *Environment and Planning D: Society and Space* 28(3): 539.

Canter, D. 1987. Implications for 'new generation' prisons of existing psychological research into prison design and use. In *Problems of Long-term Imprisonment*, edited by A. E. Bottoms and R. Light. Gower, Aldershot.

Caputo-Levine, D. D. 2013. The yard face: The contributions of inmate interpersonal violence to the carceral habitus. *Ethnography* 14(2): 165–185.

Carey, H. F. 2013. The Domestic Politics of Protecting Human Rights in Counter-Terrorism Poland's, Lithuania's, and Romania's Secret Detention Centers and Other East European Collaboration in Extraordinary Rendition. *East European Politics & Societies* 27(3): 429–465.

Carlson, K. A. 1991. What Happens and What Counts: Resident Assessments of Prison Impacts on Their Communities. *Humboldt Journal of Social Relations* 17(1&2): 211–237.

Carlton, B. and M. Segrave 2011. Women's survival post-imprisonment: Connecting imprisonment with pains past and present. *Punishment & Society* 13(5): 551–570.

Carney, M. A. 2013. Border Meals: Detention Center Feeding Practices, Migrant Subjectivity, and Questions on Trauma. *Gastronomica; The Journal of Food and Culture* 13(4): 32–46.

Carnochan, B. W. 1998. The literature of confinement. In *The Oxford History of the Prison: The Practice of Punishment in Western Society*, edited by N. Morris and D.J. Rothman. Oxford University Press, Oxford, pp. 380–406.

Casella, E. C. 2005. Prisoner of His Majesty: Postcoloniality and the archaeology of British penal transportation. *World Archaeology* 37(3): 453–467.

Cavalier, E. S. 2011. Men at Sport: Gay Men's Experiences in the Sport Workplace. *Journal of Homosexuality* 58(5): 626–646.

Cesaroni, C. and M. Peterson-Badali 2005. Young offenders in custody: Risk and adjustment, *Criminal Justice and Behavior* 32(3): 251–277.

Chantraine, G. 2005. Expériences carcérales et savoirs minoritaires. *Informations sociales* 7: 42–52.

Charlesworth, A. 1994. Contesting places of memory: The case of Auschwitz *Environment and Planning D: Society and Space* 12: 579–593.

Che, D. 2005. Constructing a Prison in the Forest: Conflicts Over Nature, Paradise and Identity. *Annals of the Association of American Geographers* 95(4): 809–831.

Cheliotis, L. K. 2010. The ambivalent consequences of visibility: Crime and prisons in the mass media. *Crime, Media, Culture* 6(2): 169–184.

Cherry, T. and Kunce, M. 2001. Do Policymakers Locate Prisons for Economic Development? *Growth and Change* 32(4): 533–547.

Cloke P., P. Milbourne and R. Widdowfield 2003. The complex mobilities of homeless people in rural England. *Geoforum* 34(1): 21–35.

Codd, H. 1998. Older women, criminal justice, and women's studies. *Women's Studies International Forum* 21(2): 183–192.

Codd, H. 2007. Prisoners' Families and Resettlement: A Critical Analysis. *The Howard Journal* 46(3): 255–263.

Cohen, S. 1985. *Visions of Social Contro. Crime, Punishment and Classification.* Polity Press, Cambridge.

Cohen, S. and L. Taylor 1972. *Psychological Survival: The Experience of Long-term Imprisonment.* Penguin, Harmondsworth.

Combessie, P. 2002. Marking the Carceral Boundary Penal Stigma in the Long Shadow of the Prison. *Ethnography* 3(4): 535–555.

Comfort, M. 2002. 'Papa's house' The prison as domestic and social satellite. *Ethnography* 3(4): 467–499.

Comfort, M. 2003. In the Tube at San Quentin: The 'Secondary Prisonization' of Women Visiting Inmates. *Journal of Contemporary Ethnography* 32(1): 77–107.

Comfort, M. 2008. *Doing Time Together: Love and Family in the Shadow of the Prison.* University of Chicago Press, Chicago.

Conlon, D. 2013. Hungering for Freedom: Asylum Seekers' Hunger Strikes – Rethinking Resistance as Counter-Conduct. In *Carceral Spaces: Mobility and Agency in Imprisonment and Migrant Detention*, edited by D. Moran, N. Gill and D. Conlon. Ashgate, Farnham, pp. 133–148.

Connell, R. 2012. Transsexual women and feminist thought. *Signs* 37: 857–881.

Cooke, P. 2000. Kilmainham Gaol: Interpreting Irish Nationalism and Republicanism. *Open Museum Journal* 2: 1–11.

Cope, N. 2003. 'It's no time or high time': Young offenders' experiences of time and drug use in prison. *Howard Journal* 42(2): 158–175.

Cosgrove, D. and S. Daniels (eds) 1988. *The Iconography of Landscape: Essays on the Symbolic Representation, Design and Use of Past Environments* (Vol. 9). Cambridge University Press, Cambridge.

Courtright, K. E., S. H. Packard, M. J. Hannan and E. T. Brennan 2010. Prisons and rural Pennsylvania communities: Exploring the health of the relationship and the possibility of improvement. *The Prison Journal* 90(1): 69–93.

Craddock, S. 2000. Disease, social identity, and risk: Rethinking the geography of AIDS. *Transactions of the Institute of British Geographers* 25(2): 153–168.

Crampton, J. 2009. Cartography, Maps 2.0. *Progress in Human Geography* 33: 91–100.

Crampton, J. W. and S. Elden (eds) 2007. *Space, Knowledge and Power: Foucault and Geography.* Ashgate, Farnham.

Crawley, E. 2005. Institutional thoughtlessness in prisons and its impacts on the day-to-day prison lives of elderly men. *Journal of Contemporary Criminal Justice* 21(4): 350–363.

Cresswell, T. 1999. Embodiment, power and the politics of mobility: The case of female tramps and hobos *Transactions of the Institute of British Geographers* 24: 175–192

Cresswell, T. 2006. *On the Move: Mobility in the Modern Western World.* Routledge, London.

Cresswell, T. 2008. Understanding mobility holistically: The case of Hurricane Katrina. In *The Ethics of Mobilities: Rethinking Place, Exclusion, Freedom and Environment*, edited by S. Bergmann and T. Sager. Ashgate, Aldershot, pp. 25–38.

Cresswell, T. 2010. Towards a politics of mobility. *Environment and Planning D: Society and Space* 28: 17–31.

Crewe, B. 2009. *The Prisoner Society: Power, Adaption, and Social Life in an English Prison.* Oxford University Press, Oxford.

Crewe, B. 2011. Depth, weight, tightness: Revisiting the pains of imprisonment. *Punishment and Society* 13(5): 509–529.

Crewe, B., J. Warr, P. Bennett and A. Smith 2014. The emotional geography of prison life. *Theoretical Criminology* 18(1): 56–74.

Criminal Justice Act 2003, The Stationery Office, London.

Crush, J. 1994. Scripting the compound: Power and space in the South African mining industry. *Environment and Planning D* 12: 301–301.

Cunningham, M., R. E. Glenn, H. M. Field and J. R. Jakobsen 1985. Dental Disease Prevalence in a Prison Population. *Journal of Public Health Dentistry* 45(1): 48–52.

Curtis, S., Gesler, W., Fabian, K., et al. 2007. Therapeutic landscapes in hospital design: A qualitative assessment by staff and service users of the design of a new mental health inpatient unit. *Environment and Planning C* 25(4): 591.

Czajka, A. 2005. Inclusive Exclusion: Citizenship and the American Prisoner and Prison. *Studies in Political Economy* 76: 111–142.

Davidson, J. and C. Milligan 2004. Embodying emotion sensing space: Introducing emotional geographies. *Social & Cultural Geography* 5(4): 523–532.

Davidson, J. and L. Bondi 2004. Spatialising affect, affecting space: An introduction. *Gender, Place and Culture* 11: 373–374.

Davidson, J., M. Smith and L. Bondi (eds) 2005. *Emotional Geographies.* Ashgate, Aldershot.

Davis, L. M. and S. Pacchiana 2004. Health Profile of the State Prison Population and Returning Offenders: Public Health Challenges. *Journal of Correctional Health Care* 10(3): 303–331.

Davis, M. 1990. *City of Quartz: Evacuating the Future in Los Angeles.* Verso, London.

Davis, M. 2006. *City of Quartz: Excavating the Future in Los Angeles* (New Edition). Verso, London.

Davison, R. L. 1931. Prison Architecture. *Annals of the American Academy of Political and Social Science* 157: 33–39.

de Certeau, M. 1984. *The Practice of Everyday Life.* University of California Press, California.

de Dardel, J. 2013. Resisting 'Bare Life': Prisoners' Agency in the New Prison Culture Era in Colombia. In *Carceral Spaces: Mobility and Agency in Imprisonment and Migrant Detention*, edited by D. Moran, N. Gill and D. Conlon. Ashgate, Farnham, pp. 183–198.

De Meis, C. 2002. House and Street: Narratives of Identity in a Liminal Space among Prostitutes in Brazil. *Ethos* 30(1/2): 3–24.

Deacon, H. 2004. Intangible Heritage in Conservation Management Planning: The Case of Robben Island. *International Journal of Heritage Studies* 10(3): 309–319.

DeLyser, D. 1999. Authenticity on the ground: Engaging the past in a California ghost town. *Annals of the Association of American Geographers* 89(4): 602–632.

Demonchy, C. 1998. Architecture et évolution du système pénitentiaire. *Les cahiers de la sécurité intérieure* 31: 79–89.

Demonchy, C. 2000. L'institution mal dans ses murs. In *La prison en changement, Ramonville-Saint-Agne,* edited by C. Veil and D. Lhuilier. Érès, Ramonville-Saint-Agne, pp. 159–184.

Demonchy, C. 2004. L'architecture des prisons modèles françaises. In *Gouverner, enfermer. La prison, un modèle indépassable?* edited by P. Artières and P. Lascoumes. Paris: Presses de Sciences Po, pp. 269–293.

deVerteuil, G., J. May and J. von Mahs 2009. Complexity not collapse: Recasting the geographies of homelessness in a 'punitive' age. *Progress in Human Geography* 33(5): 646–666.

Dewar, M. and C. Fredericksen 2003. Prison Heritage, Public History and Archaeology at Fannie Bay Gaol, Northern Australia. *International Journal of Heritage Studies* 9(1): 45–63.

Dey, E. 2009. Prison Tours as a Research Tool in the Golden Gulag. *Journal of Prisoners on Prisons* 18(1/2).

Dhlakama, G. H., O. A. Adeleke, G. A. Chukwu, et al. 2006. 'Oral Health Practices of Inmates of the Jos Main Prison', abstract from the 5th Annual Scientific Congress of IADR Nigerian Division, 29 August – 1 September 2006.

Dickinson, C. J. 2003. Group: Rural prisons don't provide economic boost. *Business Journal – Central New York* 17: 4–6.

Dirsuweit, T. 1999. Carceral spaces in South Africa: A case study of institutional power, sexuality and transgression in a women's prison. *Geoforum* 30: 71–83.

Ditchfield, J. 1990. *Control in Prisons: A Review of the Literature.* Home Office Research Study 118, HSMO, London.

Dodgshon, R. A. 2008. Geography's place in time. *Geografiska Annaler: Series B, Human Geography* 90(1): 1–15.

Doty, R. L. and E. S. Wheatley 2013. Private Detention and the Immigration Industrial Complex *International Political Sociology* 7: 426–443.

Douglas, N. and E. Plugge 2008. The health needs of imprisoned female juvenile offenders: The views of the young women prisoners and youth justice. *Professionals International Journal of Prisoner Health* 4(2): 66–76.

Downes, D. 2007. Visions of Penal Control in the Netherlands. In *Crime, Punishment, and Politics in Comparative Perspective – Crime and Justice: A Review of Research*, edited by M. Tonry. University of Chicago Press, Chicago, pp. 93–125.

Drake, D. 2012. *Prisons, Punishment and the Pursuit of Security*. Palgrave Macmillan, Basingstoke.

Drake, D. H. and R. Earle 2013. On the inside: Prison ethnography around the globe: Deborah H. Drake and Rod Earle introduce the articles in the themed section. *Criminal Justice Matters* 91(1): 12–13.

Draper, S. 2012. *Afterlives of Confinement*. University of Pittsburgh Press, Pennsylvania.

Driver, F. 1985. Power, space, and the body: A critical assessment of Foucault's Discipline and Punish. *Environment and Planning D: Society and Space* 3(4): 425–446.

Drummond, L. B. W. 2000. Street Scenes: Practices of Public and Private Space in Urban Vietnam. *Urban Studies* 37(12): 2377–2391.

Dullum, J. and T. Ugelvik (eds) 2011. *Nordic Prison Practice and Policy, Exceptional Or Not?: Exploring Penal Exceptionalism in the Nordic Context*. Routledge, London.

Duncan J. and N. Duncan 1988. (Re)reading the Landscape. *Environment and Planning D: Society and Space* 6: 117–126.

Duncan, J. S. 1990. *The City as Text: The Politics of Landscape Interpretation in the Kandyan Kingdom*. Cambridge University Press, New York.

Dunn, J. A. 1998. *Driving Forces: The Automobile, Its Enemies, and the Politics of Mobility*. Brookings Institution, Washington.

Eason, J. 2010. Mapping prison proliferation: Region, rurality, race and disadvantage in prison placement. *Social Science Research* 39(6): 1015–1028.

Eason, J. M. 2012. Extending the Hyperghetto: Toward a Theory of Punishment, Race, and Rural Disadvantage. *Journal of Poverty* 16(3): 274–295.

Elden, S. 2001. *Mapping the Present: Heidegger, Foucault and the Project of a Spatial History*. Continuum, London.

Engel, D. J. 2007. *When a Prison Comes to Town: Siting, Location and Perceived Impacts of Correctional Facilities in the MidWest*. Unpublished PhD thesis, The University of Nebraska, Lincoln.

Evans, R. 1982. *The Fabrication of Virtue: English Prison Architecture 1750–1840*. Cambridge University Press, Cambridge.

Fairweather, L. and S. McConville (eds) 2000. *Prison Architecture: Policy, Design, and Experience*. Elsevier, Oxford.

Farber, M. L. 1944. Suffering and time perspective of the prisoner. *University of Iowa Studies in Child Welfare* 20: 153–227.

Farrigan, T. L. and A. K. Glasmeier (n.d.) *The Economic Impacts of the Prison Development Boom on Persistently Poor Rural Places.* Working Paper of Poverty in America, Penn State University.

Farrington, K. 1992. The Modern Prison as Total Institution? Public Perception versus Objective Reality. *Crime & Delinquency* 38(6): 6–26.

Fenton, L. 2005. Citizenship in Private Space. *Space and Culture* 8(2): 180–192.

Ferrant, A. 1997. *Containing the Crisis: Spatial Strategies and the Scottish Prison System.* Unpublished PhD thesis, University of Edinburgh, Department of Geography.

Fiddler, M. 2010. Four walls and what lies within: The meaning of space and place in prisons. *Prison Service Journal* 187: 3–8.

Fiddler, M. 2012. Editorial Comment. *Prison Service Journal* 199: 2–4.

Fior, M. 1993. L'espace emprisonné (Imprisoned space). *Bulletin de la Société Neuchâteloise de Géographie* 37: 43–54.

Fisher, M. A., G. H. Gilbert and B. J. Shelton 2005. Effectiveness of dental services in facilitating recovery from oral disadvantage. *Quality of Life Research* 14: 197–206.

Flynn, M. K. 2011. Decision-making and contested heritage in Northern Ireland: The former Maze prison/Long Kesh. *Irish Political Studies* 26(3): 383–401.

Foley, M. and J. Lennon 2000. *Dark Tourism: The Attraction of Death and Disaster.*Cengage Learning EMEA, London.

Foucault, M. [1967] 1998. Different spaces. In *Aesthetics, Method, and Epistemology*, M. Foucault. Penguin Books, London.

Foucault, M. 1979. *Discipline and Punish: The Birth of the Prison.* Vintage, New York.

Foucault, M. 1991. Governmentality. In *The Foucault Effect: Studies in Governmentality*, edited by G. Burchell, C. Gordon and P. Miller. Harvester Wheatsheaf, London, pp. 87–104.

Foucault, M. 2003. *Abnormal: Lectures at the Collège de France 1974–75*, trans. G. Burchell. Verso, London.

Foucault, M. 2007. *Security, Territory, Population*, trans. M. Senellart, F. Ewald and A.I. Davidson. Palgrave Macmillan, New York.

Fraser, J. 2000. An American seduction: Portrait of a prison town. *Michigan Quarterly Review* 39(4): 775–795.

Freudendal-Pedersen, M. 2009. *Mobility in Daily Life: Between Freedom and Unfreedom.* Ashgate, Farnham.

Frost, N. A. 2010. Beyond public opinion polls: Punitive public sentiment and criminal justice policy. *Sociology Compass* 4(3): 156–168.

Gabel, K. and D. Johnston 1995. *Children of Incarcerated Parents.* Lexington, New York.

Garland, D. 1990. *Punishment and Modern Society: A Study in Social Theory.* University of Chicago Press, Chicago.

Garland, D. 2001. *The Culture of Control.* Oxford University Press, Oxford.

Garton-Smith, J. 2000. The Prison Wall: Interpretation Problems for Prison Museums. *Open Museum Journal* 2.

Gear, S. 2005. Rules of Engagement: Structuring Sex and Damage in Men's Prisons and Beyond. *Culture, Health and Sexuality* 7(3): 195–208.

Gear, S. 2007. Behind the Bars of Masculinity: Male Rape and Homophobia in and about South African Men's Prisons. *Sexualities* 10: 209–227.

Genocchio, B. 1995. Discourse, discontinuity, difference: The question of 'other' spaces. In *Postmodern Cities and Spaces*, edited by S. Watson and K. Gibson. Blackwell, Oxford, pp. 35–46.

Genter, S., G. Hooks and C. Mosher 2013. Prisons, jobs and privatization: The impact of prisons on employment growth in rural US counties, 1997–2004. *Social Science Research* 42(3): 596–610.

Gesler, W. M. 1992. Therapeutic landscapes: Medical issues in light of the new cultural geography. *Social Science & Medicine* 34(7): 735–746.

Gesler, W., M. Bell, S. Curtis, et al. 2004. Therapy by design: Evaluating the UK hospital building program. *Health & Place* 10(2): 117–128.

Gibson, C. 2010. Geographies of tourism: (Un)ethical encounters. *Progress in Human Geography* 34(4): 521–527.

Giddens, A. 1981. Agency, Institution and Time-Space Analysis. In *Advances in Social Theory and Methodology: Toward an Integration of Micro and Macro Sociologies*, edited by K. Knorr-Cetina and A. V. Cicourel. Routledge & Kegan Paul, Boston, pp. 161–174.

Giddens, A. 1984. *The Constitution of Society: Outline of the Theory of Structuration*. Polity Press, Cambridge.

Gill, N. 2009. Governmental mobility: The power effects of the movement of detained asylum seekers around Britain's detention estate. *Political Geography* 28(3): 186–196.

Gill, N. 2013. Mobility versus Liberty? The Punitive Uses of Movement Within and Outside Carceral Environments. In *Carceral Spaces: Mobility and Agency in Imprisonment and Migrant Detention*, edited by D. Moran, N. Gill and D. Conlon. Ashgate, Farnham, pp. 19–36.

Gill, N., D. Conlon, D. Moran and A. Burridge (forthcoming). Circuitry and Peno-Cartography: New Directions in Carceral Geography. *Progress in Human Geography.*

Gilmore, R. W. 1999. You have dislodged a boulder: Mothers and prisoners in the post Keynesian California landscape. *Transforming Anthropology* 8(1–2): 12–38.

Gilmore, R. W. 2002. Fatal Couplings of Power and Difference: Notes on Racism and Geography. *The Professional Geographer* 54: 15–24.

Gilmore, R. W. 2007. *Golden Gulag: Prisons, Surplus, Crisis, and Opposition in Globalizing California*. University of California Press, California.

Glasmeier, A. K. and T. Farrigan 2007. The Economic Impacts of the Prison Development Boom on Persistently Poor Rural Places. *International Regional Science Review* 30(3): 274–299.

Goffman, E. 1959. *The Presentation of Self in Everyday Life.* Doubleday, New York.

Goffman, E. 1961. *Asylums: Essays on the Social Situation of Mental Patients and Other Inmates.* Anchor Books, Garden City, NY.

Goffman, E. 1963. *Stigma: Notes on the Management of Spoiled Identity.* Prentice Hall, Englewood Cliffs, NJ.

Gottschalk, M. 2009. The long reach of the carceral state: The politics of crime, mass imprisonment, and penal reform in the United States and abroad. *Law & Social Inquiry* 34(2): 439–472.

Graham, B. and S. McDowell 2007. Meaning in the Maze: The heritage of Long Kesh. *Cultural Geographies* 14(3): 343–368.

Graham, C. and V. Royster 1990. Rehabilitation of a Paraplegic Prisoner: Conflicts for Patient and Nurses. *Rehabilitation Nursing* 15(4): 197–200.

Graham, H. 1984. *Women, Health and the Family.* Harvester Wheatsheaf, Brighton.

Grant, E. and P. Memmott 2007. The Case for Single Cells and Alternative Ways of viewing Custodial Accommodation for Australian Aboriginal Peoples. *Flinders Journal of Law Reform* 10: 631–647.

Greer, C. and Y. Jewkes 2005. Extremes of otherness: Media images of social exclusion. *Social Justice* 32(1): 20–31.

Gregory, D. and J. Urry (eds) 1985. *Social Relations and Spatial Structures.* Macmillan, London.

Gregory, D. 2006. The Black Flag: Guantanamo Bay and the Space of Exception. *Geografisker Annaler B* 88(4): 405–427.

Groot, B. and E. J. Latessa 2007. 'The Effects of a New Prison on the Local Community'. Report prepared through a contract with the Matanuska-Susitna Borough with the coordination of the University of Alaska, College of Health and Social Wekfare, Justice Center in order to have an independent third party address the most commonly raised concerns and issues during the Phase II site selection public hearings.

Hägerstrand, T. 1975. Space, time and human conditions. In *Dynamic Allocation of Urban Space,* edited by A. Karlqvist, L. Lundqvist and F. Snickars. Saxon House, Farnborough, pp. 3–14.

Hägerstrand, T. 1982. Diorama, path and project. *Tijdschrift voor Economische en Sociale Geografie* 73(6): 323–339.

Hall, E. 2000. Blood, Brain and Bones: Taking the Body Seriously in the Geography of Health and Impairment. *Area* 32: 21–29.

Hallsworth, S. and J. Lea 2011. Reconstructing Leviathan: Emerging contours of the security state. *Theoretical Criminology* 15(2): 141–157.

Hamermesh, D.S. and J.E. Biddle 1994. Beauty and the Labor Market. *American Economic Review* 84: 1174–1194.

Hamilton, C. 2014. Reconceptualizing Penality: Towards a Multidimensional Measure of Punitiveness. *British Journal of Criminology* 54, 321–343.

Hammett, T. M., C. Roberts and S. Kennedy 2001. Health-Related Issues in Prisoner Reentry. *Crime & Delinquency* 47(3): 390–409.

Hancock, L. 2004. Criminal Justice, Public Opinion, Fear and Popular Politics. In *The Cavendish Student Handbook of Criminal Justice and Criminology*, edited by J. Muncie and D. Wilson. Cavendish, London.

Hancock, L. and G. Mooney 2012. Beyond the penal state: Advanced marginality, social policy and anti-welfarism. In *Criminalisation and Advanced Marginality: Critically Exploring the Work of Loïc Wacquant*, edited by P. Squires and J. Lea. The Policy Press, Bristol, pp. 107–128.

Hancock, P. and Jewkes, Y. 2011. Architectures of incarceration: The spatial pains of imprisonment. *Punishment and Society* 13(5): 611–629.

Hannah, M. 1997. Space and the restructuring of disciplinary power: An interpretive review *Geografiska Annaler* 79B: 171–180.

Hannam K., M. Sheller and J. Urry 2006. Editorial: Mobilities, Immobilities and Moorings. *Mobilities* 1(1): 1–22.

Hannan, M. J. and K. E. Courtright 2011. Exploring the Perception of Economic Impact of State Correctional Institutions in Rural Pennsylvania. *International Journal of Business and Social Science* 2(22): 51–69.

Hannon, T. 2007. *Children as Unintended Victims of Legal Process: A Review of Policies and Legislation Affecting Children with Incarcerated Parents.* Victorian Association for the Care and Resettlement of Offenders, Melbourne.

Hansen, N. and C. Philo 2007. The Normality of Doing Things Differently: Bodies, Spaces and Disability Geography. *Tijdschrift voor Economische en Sociale Geografie* 98(4): 493–506.

Harcourt, B. E. 2008. 'Neoliberal Penality: The Birth of Natural Order, the Illusion of Free Markets', University of Chicago Law & Economics, Olin Working Paper No. 433, http://ssrn.com/abstract=1278067.

Harley, B. 1988. Maps, Knowledge and Power. In *The Iconography of Landscape*, edited by D. Cosgrove and S. Daniels. Cambridge University Press, Cambridge.

Hassine, V. 2004. *Life without Parole: Living in Prison Today* (3rd edn). Roxbury Press, California.

Hassine, V. 2010. *Life without Parole: Living and Dying in Prison Today* (5th edn). Oxford University Press, New York.

Hensley, C., T. Castle and R. Tewkesbury 2003. Inmate-to-Inmate Sexual Coercion in a Prison for Women. *Journal of Offender Rehabilitation* 37(2): 77–87.

Herbert, S. and E. Brown 2006. Conceptions of space and crime in the punitive neoliberal city. *Antipode* 38(4): 755–777.

Hiemstra, N. 2013. 'You don't even know where you are': Chaotic Geographies of US Migrant Detention and Deportation. In *Carceral Spaces: Mobility and Agency in Imprisonment and Migrant Detention*, edited by D. Moran, N. Gill and D. Conlon. Ashgate, Farnham.

HM Prison Service 2001. *Prison Service Order 1900 Certified Prisoner Accommodation.* HM Prison Service.

Hoelscher, S. and D. H. Alderman 2004. Memory and place: Geographies of a critical relationship. *Social & Cultural Geography* 5(3): 347–355.

Hooks, G., C. Mosher, T. Rotolo and L. Lobao 2004. The prison industry: Carceral expansion and employment in US counties 1969–1994. *Social Sciences Quarterly* 85, 37–57.

Hopkins, P. 2011. Towards critical geographies of the university campus: Understanding the contested experiences of Muslim students. *Transactions of the Institute of British Geographers* 36(1): 157–169.

Horster, M. 2004. The trade in political prisoners between the two German states, 1962–89. *Journal of Contemporary History* 39(3): 403–424.

Houston, J. G., D. C. Gibbons and J. F. Jones 1988. Physical Environment and Jail Social Climate. *Crime & Delinquency* 34(4): 449–466.

Howitt, R. 2001. Frontiers, Borders, Edges: Liminal Challenges to the Hegemony of Exclusion. *Australian Geographical Studies* 39(2): 233–245.

Howse, K. 2003. *Growing Old in Prison*. Prison Reform Trust, London.

Hoyman, M. 2002. Prisons in North Carolina: Are they a Viable Strategy for Rural Communities? *International Journal of Economic Development* 4(1).

Huling, T. 2002. Building a prison economy in rural America. In Chesney-Lind, M. and M. Mauer (eds) *Invisible Punishment: The Collateral Consequences of Mass Imprisonment*. The New Press, New York, pp. 197–213.

Hurd, D. 2000. *Memoirs*. Little Brown, London.

Husserl, E. 1983. *Ideas Pertaining to a Pure Phenomenology and to a Phenomonological Philosophy: General Introduction to a Pure Phenomenology*. Kluwer Academic, Dordrecht.

Huxley, M. 2006. Spatial rationalities: Order, environment, evolution and government. *Social & Cultural Geography* 7(5): 771–787.

Hyndman, J. 2012. The geopolitics of migration and mobility. *Geopolitics* 17(2): 243–255.

Indra, D. 1998. *Engendering Forced Migration*. Berghahn, Oxford.

Jackson, N. 1999. The Council Tenants' Forum: A Liminal Public Space between Lifeworld and System? *Urban Studies* 36(1): 43–58.

Jackson, P. 1998. Domesticating the street: The contested spaces of the high street and the mall. In *Images of the Street: Planning, Identity and Control in Public Space*, edited by N. R. Fyfe. Routledge, London, pp. 176–191.

Jacobs, D. and R. E. Helms 1996. Toward a political model of incarceration: A time-series examination of multiple explanations for prison admission rates. *American Journal of Sociology* 102(2): 323–357.

Jacobs, J. M. 2006. A geography of big things. *Cultural Geographies* 13(1): 1–27.

Jacobs, J. M. and P. Merriman 2011. Practising architectures. *Social & Cultural Geography* 12(3): 211–222.

Jargin, S. V. 2009. Some aspects of dental care in Russia. *Indian Journal of Dental Research* 20: 518–519.

Jefferson, B. J. 2014. 'When Public Space is the Prison: NYPD, Hyperpolicing and Geographies of Crime Control', paper presented at the 2014 Annual Meeting of the Association of American Geographers, Tampa, FL, 8–12 April 2014.

Jenness, V. and S. Fenstermaker 2013. Agnes Goes to Prison: Gender Authenticity, Transgender Inmates in Prisons for Men, and Pursuit of 'The Real Deal'. *Gender & Society*, doi: 10.1177/0891243213499446.

Jensen, A. 2011. Mobility, space and power: On the multiplicities of seeing mobility. *Mobilities* 6(2): 255–271.

Jewkes, Y. 2002. *Captive Audience: Media, Masculinity and Power in Prisons.* Routledge, London.

Jewkes, Y. 2005. Loss, liminality and the life sentence: Managing identity through a disrupted lifecourse. In *The Effects of Imprisonment*, edited by A. Liebling and S. Maruna. Willan, Cullompton.

Jewkes, Y. 2013a. On Carceral Space and Agency. In *Carceral Spaces: Mobility and Agency in Imprisonment and Migrant Detention*, edited by D. Moran, N. Gill and D. Conlon. Ashgate, Farnham, pp. 127–132.

Jewkes, Y. 2013b. The Aesthetics and Anaesthetics of Prison Architecture. In *Architecture and Justice: Judicial Meanings in the Public Realm*, edited by J. Simon, N. Temple and R. Tobe, Ashgate, Farnham, pp. 9–22.

Jewkes, Y. and D. Moran 2014. Should Prison Architecture be Brutal, Bland or Beautiful? *Scottish Justice Matters* 2(1): 8–11.

Jewkes, Y. and H. Johnston 2007. The evolution of prison architecture. In *Handbook on Prisons*, edited by Y. Jewkes. Willan, Cullompton.

Johnson, L. C. 2008. Re-placing gender? Reflections on 15 years of Gender, Place and Culture. *Gender, Place and Culture* 15(6): 561–574.

Johnston, N. 2000. *Forms of Constraint: A History of Prison Architecture.* University of Illinois Press, Illinois.

Jones, M. 2009. Phase space: Geography, relational thinking, and beyond. *Progress in Human Geography* 33(4): 487–506.

Kantrowitz, N. 1996. *Close Control: Managing a Maximum Security Prison: The Story of Ragen's Stateville Penitentiary.* Harrow and Heston Publishers, Albany, NY.

Kaplan-Lyman, J. 2012. A Punitive Bind: Policing, Poverty and Neoliberalism in New York City. *Yale Human Rights and Development Journal* 15(1): 177–221.

Katz, M. B. 1996. *In The Shadow of the Poorhouse: A Social History of Welfare in America New York.* Basic Books, New York.

Kaufmann, V. 2002. *Re-thinking Mobility.* Ashgate, Aldershot.

Kay, R. 2000. *Russian Women and their Organisations.* Macmillan, Basingstoke.

Kearon, A. T. 2012. Alternative Representation of the Prison and Imprisonment: Comparing Dominant Narratives in the News Media and in Popular Fictional Texts. *Prison Service Journal* 199: 4–10.

Kim, K. 2007. Psychological Coercion in the Context of Modern-Day Involuntary Labor: Revisiting United States v. Kozminski and Understanding Human Trafficking. *University of Toledo Law Review* 38: 941–972.

Kindynis, T. 2014. Ripping up the Map: Criminology and Cartography Revisited. *British Journal of Criminology* 54: 222–243.

King, R. S., M. Mauer and T. Huling 2004. An analysis of the economics of prison siting in rural communities. *Criminology and Public Policy* 3(3): 453–480.

Knie, A. 1997. Eigenzeit und Eigenraum: Zur Dialektik von Mobilitat und Verkehr. *Soziale Welt* 48(1): 39–54.

Kofman, E. 2002. Contemporary European Migrations: Civic stratification and citizenship. *Political Geography* 2(8): 1035–1054.

Koskela, H. 2000. 'The gaze without eyes': Video-surveillance and the changing nature of urban space. *Progress in Human Geography* 24(2): 243–265.

Kraftl, P. 2010. Geographies of architecture: The multiple lives of buildings. *Geography Compass* 4(5): 402–415.

Kraftl, P. and P. Adey 2008. Architecture/affect/inhabitation: Geographies of being-in buildings. *Annals of the Association of American Geographers* 98(1): 213–231.

Krames, L. and G. L. Flett 2002. The Perceived Characteristics of Holding Cell Environments: Report of a Pilot Study. *Canadian Police Research Centre.*

Kumar, K. and E. Makarova 2008. The Portable Home: The Domestication of Public Space. *Sociological Theory* 26(4): 324–343.

Kuusi, H. 2008. Prison Experiences and Socialist Sculptures – Tourism and the Soviet Past in the Baltic States. In *Touring the Past. Uses of History in Tourism. Discussion and Working Papers No. 6*, edited by A. Kostiainen and T. Syrjämaa. The Finnish University Network for Tourism Studies (FUNTS) Savonlinna, pp. 105–122.

Lamarre, J. 2001. La territorialisation de l'espace carcéral (Territorial processes in jail). *Géographie et cultures* 40(1): 77–92.

Lappi-Seppälä, T. 2002. Penal Policy and Incarceration Rates. *Corrections Today* 30–33.

Larsen, J., J. Urry and K. Axhausen 2007. *Mobilities, Networks, Geographies.* Ashgate, Aldershot.

Laurenson, P. and D. Collins 2007. Beyond punitive regulation? New Zealand local governments' responses to homelessness. *Antipode*, 39(4): 649–667.

Laws, J. 2009. Reworking therapeutic landscapes: The spatiality of an 'alternative' self-help group. *Social Science & Medicine* 69(12): 1827–1833.

Leech, M. 2005. *The Prisons Handbook.* MLA Press, Manchester.

Lefèbvre, H. 1991. *The Production of Space.* Oxford: Blackwell.

Levy, C. 2010. Refugees, Europe, camps/state of exception: 'Into the zone', the European Union and extraterritorial processing of migrants, refugees and asylum-seekers (theories and practices). *Refugee Studies Quarterly* 29(1): 92–119.

Levy, M. H., S. Quilty, L. C. Young, et al. 2003. Pox in the docks: Varicella outbreak in an Australian prison system. *Public Health* 117: 446–451.

Li, F., S. Papagiannidis and M. Bourlakis 2010. Living in 'multiple spaces': Extending our socioeconomic environment through virtual worlds. *Environment and Planning D: Society and Space* 28(3): 425–446.

Liebling, A. 2002. *Suicides in Prison.* Routledge, London.

Liebling, A. with H. Arnold 2004. *Prisons and their Moral Performance: A Study of Values, Quality, and Prison Life.* Oxford University Press, Oxford.

Locker, D. 2009. Self-Esteem and Socioeconomic Disparities in Self-Perceived Oral Health. *Journal of Public Health Dentistry* 69(1): 1–8.

Lockyer, K. 2013. *Future Prisons: A Radical Plan to Reform the Prison Estate.* Policy Exchange, London.

Longhurst, R. 2005. Fat bodies: Developing geographical research agendas. *Progress in Human Geography* 29(3): 247–259.

Lopoo, L.M. and B. Western 2005. Incarceration and the Formation and Stability of Marital Unions. *Journal of Marriage and Family* 67: 721–734.

Lorimer, H. 2008. Cultural geography: Non-representational conditions and concerns. *Progress in Human Geography* 32: 551–559.

Loyd, J., M. Mitchelson and A. Burridge (eds) 2012. *Beyond Walls and Cages: Prisons, Borders, and Global Crisis.* University of Georgia Press, Athens.

Lynch, M. 2011. Mass incarceration, legal change, and locale. *Criminology & Public Policy* 10(3): 673–698.

Lyons, G., J. Jain and D. Holley 2006. The use of travel time by rail passengers in Great Britain. *Transportation Research Part A* 41: 107–120.

Macgregor, D. M. and J. W. Balding 2005. Self-esteem as a predictor of toothbrushing behaviour in young adolescents. *Journal of Clinical Periodontology* 18(5): 312–316.

MacKain, S. J. and C. E. Messer 2004. Ending the Inmate Shuffle: An Intermediate Care Program for Inmates with a Chronic Mental Illness. *Journal of Forensic Psychology Practice* 4(2): 87–100.

Madge, C. and H. O'Connor 2005. Mothers in the making? Exploring liminality in cyber/space. *Transactions of the Institute of British Geographers* 30: 83–97.

Mahon-Daly, P. and G. J. Andrews 2002. Liminality and breastfeeding: Women negotiating space and two bodies. *Health & Place* 8: 61–76.

Mali, J. 2008. Comparison of the characteristics of homes for older people in Slovenia with Goffman's concept of the total institution. *European Journal of Social Work* 11(4): 431–443.

Mallik-Kane, K. 2005. *Returning Home Illinois Policy Brief: Health and Prisoner Reentry.* Urban Institute Justice Policy Center, Washington, DC.

Manderscheid, K. 2009. Unequal mobilities. In *Mobilities and Inequality*, edited by T. Ohnmacht, H. Maksim, and M. M. Bergman. Ashgate, Aldershot, pp. 27–50.

Marshall, A. 1997. 'Always Greener on the Other Side of the Fence? Examining the Relationship between the Built Environment, Regimes and Control in Medium Security Prisons in England and Wales'. Unpublished PhD thesis, University of Birmingham, School of Geography.

Marshall, S. 2000. Designing control and controlling for design: Towards a prison plan classification for England and Wales? In *Landscapes of Defence*, edited by J. R. Gold and G. Revill. Prentice Hall, Harlow, pp. 227–245.

Martin, D. and P. Wilcox 2012. Women, welfare and the carceral state. In *Criminalisation and Advanced Marginality: Critically Exploring the Work of Loïc Wacquant*, edited by P. Squires and J. Lea. The Policy Press, Bristol, pp. 151–170.

Martin, L. L. and M. L. Mitchelson 2009. Geographies of detention and imprisonment: Interrogating spatial practices of confinement, discipline, law, and state power. *Geography Compass* 3(1): 459–477.

Martin, R. 2000. Community perceptions about prison construction: Why not in my backyard? *The Prison Journal* 80(3): 265–294.

Maruna, S. 2001. *Making Good: How Ex-Convicts Reform and Rebuild their Lives*. American Psychological Association, Washington, DC.

Mason, P. (ed.) 2013. *Captured by the Media*. Willan, Cullompton.

Massey, D. 1993. Power-Geometry and a Progressive Sense of Place. In *Mapping the Futures: Local Cultures, Global Change*, edited by J. Bird, B. Curtis, T. Putnam, et al. Routledge, New York, pp. 59–69.

Massey, D. 1994. *Space, Place and Gender*. Polity Press, Cambridge.

Massey, D. 1995. The conceptualization of place. In *A Place in the World? Places, Cultures and Globalization*, edited by D. Massey and P. Jess. Oxford University Press, New York, pp. 45–85.

Massey, D. 1999. Space-time, 'science' and the relationship between physical geography and human geography. *Transactions of the Institute of British Geographers* 24(3): 261–276.

Massey, D. 2005. *For Space*. Sage, London.

Mathiesen, T. 2000. *Prison on Trial* (2nd edn). Waterside Press, Winchester.

Maxwell-Stewart, H. 2013. 'The Lottery of Life': Convict Tourism at Port Arthur Historic Site, Australia. *Prison Service Journal* 24–28.

May, J. and N. Thrift 2001. Introduction. In *Timespace: Geographies of Temporality*, edited by J. May and N. Thrift. Routledge, London, pp. 1–46.

McAlister, S., P. Scraton and D. Haydon 2009. *Childhood in Transition: Experiencing Marginalisation and Conflict in Northern Ireland*. Queen's University Belfast, Belfast.

McAtackney, L. 2005. 'The Negotiation of Identity at Shared Sites: Long Kesh/ Maze Prison Site, Northern Ireland', paper presented at the Forum UNESCO University and Heritage 10th International Seminar 'Cultural Landscapes in the 21st Century'. Newcastle upon Tyne, 11–16 April 2005.

McAtackney, L. 2013. Dealing with Difficult Pasts: The Dark Heritage of Political Prisons in Transitional Northern Ireland and South Africa. *Prison Service Journal* 210: 17–23.

McCorkle, R. C. 1992. Personal precautions to violence in prison. *Criminal Justice and Behavior* 19(2): 160–173.

McCormack, D. P. 2008. Engineering affective atmospheres on the moving geographies of the 1897 Andrée expedition. *Cultural Geographies* 15: 413–430.

McDowell, L. 1999. *Gender, Identity and Place. Understanding Feminist Geographies*. Cambridge University Press, Cambridge.

McGrath, C. 2002. Oral health behind bars: A study of oral disease and its impact on the life quality of an older prison population. *Gerodontology* 19(2): 109–114.

McWatters, M. 2013. Poetic Testimonies of Incarceration: Towards a Vision of Prison as Manifold Space. In *Carceral Spaces: Mobility and Agency in Imprisonment and Migrant Detention*, edited by D. Moran, N. Gill and D. Conlon. Ashgate, Farnham, pp. 199–218.

Medlicott, D. 1999. Surviving in the time machine: Suicidal prisoners and the pains of prison time. *Time and Society* 8(2): 211–230.

Medlicott, D. 2001. *Surviving the Prison Place: Narratives of Suicidal Prisoners*. Ashgate, Aldershot.

Medlicott, D. 2008. Women in prison. In *Dictionary of Prisons and Punishment*, edited by Y. Jewkes and J. Bennett. Routledge, London.

Merleau-Ponty, M. 1962. *Phenomenology of Perception*. Routledge & Kegan Paul, London.

Merriman, P. 2012. Human geography without time-space. *Transactions of the Institute of British Geographers* 37(1): 13–27.

Merriman, P., M. Jones, G. Olsson, et al. 2012. Space and Spatiality in theory. *Dialogues in Human Geography* 2: 3–22.

Messineo, F. 2009. 'Extraordinary Renditions' and State Obligations to Criminalize and Prosecute Torture in the Light of the Abu Omar Case in Italy. *Journal of International Criminal Justice* 7(5): 1023–1044.

Middleton, J. 2009. 'Stepping in time': Walking, time, and space in the city. *Environment and Planning A* 41(8): 1943–1961.

Milhaud, O. 2009a. *Séparer et punir. Les prisons françaises: mise à distance et punition par l'espace*. Doctoral dissertation, Université Michel de Montaigne-Bordeaux III.

Milhaud, O. 2009b. La clôture suffit-elle à faire un espace d'enfermement? Spatialités contradictoires et poreuses des prisons françaises contemporaines. *Espaces D'Enfermenent, Espaces Clos*, 45.

Milhaud, O. and D. Moran 2013. Penal Space and Privacy in French and Russian Prisons. In *Carceral Spaces: Mobility and Agency in Imprisonment and Migrant Detention*, edited by D. Moran, N. Gill and D. Conlon. Ashgate, Farnham, pp. 167–182.

Milligan, C., A. Gatrell and A. Bingley 2004. 'Cultivating health': Therapeutic landscapes and older people in northern England. *Social Science & Medicine* 58(9): 1781–1793.

Minca, C. 2006. Giorgio Agamben and the New Biopolitical. *Nomos Geografiska Annaler B* 4: 387–403.

Ministry of Justice of Finland 1975. *Statute on Prison Administration*. Ministry of Justice, Helsinki.

Mitchell, D. 2003. *The Right to the City: Social Justice and the Fight for Public Space*. Guilford Press, New York.

Mitchelson, M. L. 2012. Research Note – The Urban Geography of Prisons: Mapping the City's 'Other' Gated Community. *Urban Geography* 33(1): 147–157.

Mitchelson, M. L. 2013. Up the River (from Home): Where Does the Prisoner 'Count' on Census Day? In *Carceral Spaces: Mobility and Agency in*

Imprisonment and Migrant Detention, edited by D. Moran, N. Gill and D. Conlon. Ashgate, Farnham, 77–92.

Mixson, J. M., H. C. Eplee, P. H. Fell, et al. 1990. Oral Health Status of a Federal Prison Population. *Journal of Public Health Dentistry* 50(4): 257–261.

Moore, D. and K. Hannah-Moffat 2011. The liberal veil: Revisiting Canadian penality. In *The New Punitiveness. Trends, Theories, Perspectives*, edited by J. Pratt, D. Brown, M. Brown, et al. Routledge, Oxford, 85–100.

Moran, D. 2004. Exile in the Soviet forest: 'Special settlers' in northern Perm' Oblast. *Journal of Historical Geography* 30(2): 395–413.

Moran, D. 2012. Prisoner Reintegration and the Stigma of Prison Time Inscribed on the Body. *Punishment & Society* 14: 564–583.

Moran, D. 2013a. Between outside and inside? Prison visiting rooms as liminal carceral spaces. *GeoJournal* 78(2): 339–351.

Moran, D. 2013b. Carceral geography and the spatialities of prison visiting: Visitation, recidivism, and hyperincarceration. *Environment and Planning D: Society and Space* 31: 174–190.

Moran, D. 2014. Leaving behind the 'total institution'? Teeth, TransCarceral spaces and (re)inscription of the formerly incarcerated body. *Gender, Place and Culture* 21(1): 35–51.

Moran, D. and A. Keinänen 2012. The 'Inside' and 'Outside' of Prisons: Carceral Geography and Home Visits for Prisoners in Finland. *Fennia: International Journal of Geography* 190(2): 62–76.

Moran, D. and Y. Jewkes (forthcoming) 2014. 'Green' Prisons: Rethinking the 'Sustainability' of the Carceral Estate. *Social Geography.*

Moran, D. and Y. Jewkes (forthcoming). Linking the carceral and the punitive state: Researching prison architecture, design, technology and the lived experience of carceral space. *Annales de Geographie.*

Moran, D., J. Pallot and L. Piacentini 2009. Lipstick, Lace and Longing: Constructions of Femininity inside a Russian Prison. *Environment and Planning D: Society and Space* 27(4): 700–720.

Moran, D., J. Pallot and L. Piacentini 2011. The Geography of Crime and Punishment in the Russian Federation. *Eurasian Geography and Economics* 52(1): 79–104.

Moran, D., J. Pallot and L. Piacentini 2013c. Privacy in penal space: Women's imprisonment in Russia. *Geoforum* 47: 138–146.

Moran, D., L. Piacentini and J. Pallot 2012. Disciplined Mobility and Carceral Geography: Prisoner Transport in Russia. *Transactions of the Institute of British Geographers* 37: 446–460.

Moran, D., L. Piacentini and J. Pallot 2013b. Liminal TransCarceral Space: Prisoner Transportation for Women in the Russian Federation. In *Carceral Spaces: Mobility and Agency in Imprisonment and Migrant Detention*, edited by D. Moran, N. Gill and D. Conlon. Ashgate, Farnham, pp. 109–125.

Moran, D., N. Gill and D. Conlon (eds) 2013a. *Carceral Spaces: Mobility and Agency in Imprisonment and Migrant Detention*. Ashgate, Farnham.

Morin, K. M. 2013. 'Security Here is Not Safe': Violence, Punishment, and Space in the Contemporary US Penitentiary. *Environment and Planning D: Society and Space* 31(3): 381–399.

Morris, R. G. and J. L. Worrall 2010. Prison Architecture and Inmate Misconduct: A Multilevel Assessment. *Crime & Delinquency*, doi: 10.1177/001112871038 6204.

Mountz, A. 2011. The enforcement archipelago: Detention, haunting, and asylum on islands. *Political Geography* 30(3): 118–128.

Myers, D. L. and R. Martin 2004. Community member reactions to prison siting: Perceptions of prison impact on economic factors. *Criminal Justice Review* 29(1): 115–144.

Naidoo, S. 2004. 'Oral health status of prison inmates in the Western Cape', abstract from the XXXVIII Scientific Meeting of the South African Division of IADR, 2–3 September 2004.

National Audit Office 2013. *Managing the Prison Estate.* The Stationery Office, London.

Neill, W. J. 2006. Return to Titanic and lost in the Maze: The search for representation of 'post-conflict' Belfast. *Space and Polity* 10(2): 109–120.

Noga, A. 1991. Battered Wives: The Home as a Total Institution. *Violence and Victims* 6(2): 137–149.

Norton, J. 2014. 'Little Siberia, Star of the North: The Political Economy of Prison Dreams in the Adirondacks', paper presented at the 2014 Annual Meeting of the Association of American Geographers, Tampa, FL.

Nowakowski, K. 2013. Landscapes of Toxic Exclusion: Inmate Labour and Electronics Recycling in the United States. In *Carceral Spaces: Mobility and Agency in Imprisonment and Migrant Detention*, edited by D. Moran, N. Gill and D. Conlon. Ashgate, Farnham, pp. 93–108.

Offender Information Services (OIS) 2008. *Prison Technology Strategy* (version 0.8). NOMS, London.

Ojakangas, M. 2005. Impossible Dialogue on Bio-power: Agamben and Foucault. *Foucault Studies* 2: 5–28.

Oleinik, A.N. 2003. *Organized Crime, Prison, and Post-Soviet Societies.* Ashgate, Aldershot.

Ong, C.-E., C. Minca and J. Sidaway 2012. 'The Empire and its Hotel: The Changing Biopolitics of Hotel Lloyd, Amsterdam, the Netherlands', paper presented at the Royal Geographical Society – Institute of British Geographers Annual Conference, Edinburgh, UK, July 2012.

Osborn, M., T. Butler and P.D Barnard 2008. Oral health status of prison inmates – New South Wales, Australia. *Australian Dental Journal* 48(1): 34–38.

Packer, J. 2003. Disciplining Mobility: Governing and Safety. In *Foucault, Cultural Studies, and Governmentality*, edited by J. Bratich, Z. J. Packer and C. McCarthy. State University of New York Press, New York, pp. 135–163.

Palenberg, J. C. 1982. Status of Intra-German Traffic in Political Prisoners under International Law. *The California Western International Law Journal* 12: 243–286.

Pallot, J. 2007. 'Gde muzh, tam zhena' (where the husband is, so is the wife): Space and gender in post-Soviet patterns of penality. *Environment and Planning A* 39: 570–589.

Pallot, J. and L. Piacentini with D. Moran 2012. *Gender, Geography, and Punishment: The Experience of Women in Carceral Russia.* Oxford University Press, Oxford.

Pallot, J., L. Piacentini and D. Moran 2010. Patriotic Discourses in Russia's Penal Peripheries: Remembering the Mordovian Gulag. *Europe-Asia Studies* 62(1): 1–33.

Parker, N. and N. Vaughan-Williams 2009. Lines in the Sand? Towards an Agenda for Critical Border Studies. *Geopolitics* 14(3): 582–587.

Peck, J. 2003. Geography and Public Policy: Mapping the Penal State. *Progress in Human Geography* 27: 222–32.

Peck, J. and N. Theodore 2009. Carceral Chicago: Making the ex-offender employability crisis International. *Journal of Urban and Regional Research* 32(2): 251–281.

Perkins, C. and M. Dodge 2009. Satellite imagery and the spectacle of secret spaces. *Geoforum* 40(4): 546–560.

Perkinson, R. 1994. Shackled justice: Florence Federal Penitentiary and the new politics of punishment. *Social Justice* 21(3): 117–132.

Petersilia, J. 2001. Prisoner Reentry: Public Safety and Reintegration Challenges. *The Prison Journal* 81(3): 360–375.

Pettit, B. and B. Western 2004. Mass imprisonment and the life course: Race and class inequality in U.S. incarceration. *American Sociological Review* 69(2): 151–169.

Phaswana-Mafuya, N. and N. Haydam 2005. Tourists' expectations and perceptions of the Robben Island Museum – a world heritage site. *Museum Management and Curatorship* 20: 149–169.

Phillipps, M.J. 1990. Damaged Goods: Oral Narratives of the Experience of Disability in American Culture. *Social Science and Medicine* 30(8): 849–57.

Philo, C. 2001. Accumulating populations: Bodies, Institutions and Space. *International Journal of Population Geography* 7(6): 473–490.

Philo, C. 2012. Security of Geography/Geography of Security. *Transactions of the Institute of British Geographers* 37: 1–7.

Piacentini, L., J. Pallot and D. Moran 2009. Welcome to 'Malaya Rodina' (Little Homeland): Gender, Control and Penal Order in a Russian prison. *The Journal of Socio-Legal Studies* 18(4): 523–542.

Piché, J. and K. Walby 2009. Dialogue on the Status of Prison Ethnography and Carceral Tours: An Introduction. *Journal of Prisoners on Prisons* 18: 88–90.

Piché, J. and K. Walby 2010. Problematizing carceral tours. *British Journal of Criminology* 50(3): 570–581.

Piché, J. and K. Walby 2012. Carceral Tours and the Need for Reflexivity: A Response to Wilson, Spina and Canaan. *The Howard Journal of Criminal Justice* 51(4): 411–418.

Pickering, S. 2014. Floating carceral spaces: Border enforcement and gender on the high seas *Punishment & Society* 16(2): 187–205.

Pile, S. 2010. Emotions and affect in recent human geography. *Transactions of the Institute of British Geographers* 35(1): 5–20.

Pratt, G. 2007. Abandoned Women and Spaces of Exception. *Antipode* 37(5): 1052–1078.

Pratt, J. and A. Eriksson, 2012. *Contrasts in Punishment: An Explanation of Anglophone Excess and Nordic Exceptionalism.* Routledge, London.

Pratt, J., D. Brown, M. Brown, et al. (eds) 2011. *The New Punitiveness. Trends, Theories, Perspectives.* Routledge, Oxford.

Price, B. E. and R. Schwester 2010. Economic Development Subsidies and the Funding of Private Prisons. *International Journal of Public Administration* 33(3): 109–115.

Pritchard, A. and N. Morgan 2005. Hotel Babylon? Exploring hotels as liminal sites of transition and transgression. *Tourism Management* 27: 762–772.

Ramirez, M. D. 2013. Punitive Sentiment *Criminology* 51(2): 329–364.

Reid-Henry, S. 2007. Exceptional Sovereignty? Guantánamo Bay and the Re-Colonial Present *Antipode* 39(4): 627–648.

Richardson, T. and O. B. Jensen 2008. How Mobility Systems Produce Inequality: Making Mobile Subject types on the Bangkok Sky Train. *Built Environment* 34(2): 218–231.

Ricordeau, G. and O. Milhaud 2012. Prisons. Espaces du sexe et sexualisation des espaces. *Géographie et cultures* 83: 69–85.

Rieger, A. B. 2005. Der Papst ist ein weltlicher Priester. Interview with Giorgio Agamben. *Literaturen* 21–25.

Rikard, R. V. and E. Rosenberg 2007. Aging inmates: A convergence of trends in the American criminal justice system. *Journal of Correctional Health Care* 13(3): 150–162.

Robertson, I. and P. Richards (eds). 2003. *Studying Cultural Landscapes.* Arnold, London.

Robinson, M. 2001. Tourism encounters: Inter- and intra-cultural conflicts and the world's largest industry. In *Consuming Tradition, Manufacturing Heritage,* edited by N. Alsayyad. Routledge, New York, p. 67.

Rose, G., M. Degen and B. Basdas 2010. More on 'big things': Building events and feelings. *Transactions of the Institute of British Geographers* 35: 334–349.

Rosenbloom, T. 2011. Traffic light compliance by civilians, soldiers and military officers. *Accident Analysis and Prevention* 43(6): 2010–2014.

Royal Institute of Chartered Surveyors (RICS) *Modus: The Security Issue,* 22, 11/2012.

Ruetalo, V. 2008. From Penal Institution to Shopping Mecca: The Economics of memory and the case of Punta Carretas. *Cultural Critique* 68: 38–65.

Ryan, L. and E. Chard 2013. Work in the Prison Exhibition. *Prison Service Journal* 210: 34–38.

Rygiel, K. 2011. Bordering Solidarities: Migrant activism and the politics of movement and camps at Calais. *Citizenship Studies* 15(1): 1–19.

Sager, T. 2006. Freedom as Mobility: Implications of the Distinction between Actual and Potential travelling. *Mobilities* 1(3): 465–488.

Salive, M. E., J. M. Carolla and T. Fordham Brewer 1989. Dental Health of Male Inmates in a State Prison System. *Journal of Public Health Dentistry* 49(2): 83–86.

Sampson, R. J. and J. Laub 1993. *Crime in the Making: Pathways and Turning Points through Life.* Harvard University Press, Cambridge, MA.

Schaeffer, M. A., A. Baum, P. B. Paulus and G. G. Gaes 1988. Architecturally Mediated Effects of Social Density in Prison. *Environment and Behavior* 20(1): 3–20.

Schichor, D. 1992. Myths and Realities in Prison Siting. *Crime & Delinquency* 38(1): 70–87.

Schliehe, A. K. 2014. Inside 'the Carceral': Girls and Young Women in the Scottish Criminal Justice System. *Scottish Geographical Journal* 130(2): 71–85.

Schuster, L. 2005. The Continuing Mobility of Migrants in Italy: Shifting between Places and Statuses. *Journal of Ethnic and Migration Studies* 31(4): 757–774.

Schwanen, T. 2007. Matter(s) of interest: Artefacts, spacing and timing. *Geografiska Annaler: Series B, Human Geography* 89(1): 9–22.

Schwanen, T., M.-P. Kwan and F. Ren 2008. How fixed is fixed? Gendered rigidity of space–time constraints and geographies of everyday activities. *Geoforum* 39(6): 2109–2121.

Schwartz, B. 1972. Deprivation of privacy as a 'functional prerequisite': The case of the Prison. *The Journal of Criminal Law, Criminology and Police Science* 63(2): 229–239

Scraton, P. and L. Moore 2005. Degradation, Harm and Survival in a Women's Prison. *Social Policy & Society* 5(1): 67–78.

Sechrest, D. K. 1992. Locating prisons: Open versus closed approaches to siting. *Crime & Delinquency* 38(1): 88–104.

Shabazz, R. 2009. 'So High You Can't Get Over it, So Low You Can't Get Under it': Carceral Spatiality and Black Masculinities in the United States and South Africa. *Souls* 11(3): 276–294.

Shackley, M. 2001: Potential Futures for Robben Island: Shrine, museum or theme park? *International Journal of Heritage Studies* 7(4): 355–363.

Shammas, V. L. 2014. The pains of freedom: Assessing the ambiguity of Scandinavian penal exceptionalism on Norway's Prison Island. *Punishment & Society* 16(1): 104–123.

Shanks, G. D., S. I. Hay and D. J. Bradley 2008. Malaria's indirect contribution to all-cause mortality in the Andaman Islands during the colonial era. *The Lancet Infectious Diseases* 8(9): 564–570.

Sharkey, L. 2010. Does Overcrowding in Prisons Exacerbate the Risk of Suicide among Women Prisoners? *The Howard Journal* 49(2): 111–124.

Sheller, M. and J. Urry 2006. The new mobilities paradigm. *Environment and Planning A* 38: 207–226.

Shields, R. 2003. *The Virtual*. Routledge, London.

Sibley, D. and B. Van Hoven 2008. The contamination of personal space: Boundary construction in a prison environment. *Area* 41(2): 198–206.

Sidaway, J. D. 2010. 'One Island, One Team, One Mission': Geopolitics, Sovereignty, 'Race' and Rendition. *Political Geography* 15(4): 667–683.

Silvey, R. 2004. Power, difference and mobility: Feminist advances in migration studies. *Progress in Human Geography* 28: 490–506.

Simon, J. 2007. *Governing through Crime: How the War on Crime Transformed American Democracy and Created a Culture of Fear*. Oxford University Press, Oxford.

Simon, J., N. Temple and R. Tobe (eds) 2012. *Architecture and Justice*. Ashgate, Farnham.

Simonsen, K. 2013. In quest of a new humanism. Embodiment, experience and phenomenology as critical geography. *Progress in Human Geography* 37(1): 10–26.

Siserman, C. 2012. *Reconsidering the Environmental Space of Prisons: A Step Further towards Criminal Reform*. GRIN Verlag, Munich.

Skeggs, B. 2004. *Class, Self, Culture*. Routledge, London.

Skorpen, A., N. Anderssen, C. Oeye and A.K. Bjelland 2008. The smoking-room as psychiatric patients' sanctuary: A place for resistance. *Journal of Psychiatric and Mental Health Nursing* 15(9): 728–736.

Sloan, J. 2012. 'You Can See Your Face in My Floor': Examining the Function of Cleanliness in an Adult Male Prison. *The Howard Journal of Criminal Justice* 51(4): 400–410.

Smith, A. and K. Buckley 2007. Convict landscapes: Shared heritage in New Caledonia. *Historic Environment* 20(2): 27–31.

Smith, C. 2002. Punishment and pleasure: Women, food and the imprisoned body. *The Sociological Review* 50(2): 197–214.

Smith, C. 2013. Spaces of Punitive Violence. *Criticism* 55(1): 161–168.

Smith, H. P. 2013. Reinforcing Experiential Learning in Criminology: Definitions, Rationales, and Missed Opportunities Concerning Prison Tours in the United States. *Journal of Criminal Justice Education* 24(1): 50–67.

Smoyer, A. and K. M. Blankenship 2013. Dealing food: Female drug users' narratives about food in a prison place and implications for their health. *International Journal of Drug Policy*, http://dx.doi.org/doi:10.1016/j.drugpo.2013.10.013.

Snacken, S. 2010. Resisting punitiveness in Europe? *Theoretical Criminology* 14(3): 273–292.

Snoek, A. 2010. Agamben's Foucault: An overview. *Foucault Studies* 10: 44–67.

Soja, E. 1985. Regions in context: Spatiality, periodicity, and the historical geography of the regional question. *Environment and Planning D: Society and Space* 3: 175–190.

Soja, E. W. 1989. *Postmodern Geographies. The Reassertion of Space in Critical Social Theory.* Verso, London.

Solzhenitsyn, A. 1974. *The Gulag Archipelago.* Collins, Glasgow.

Sörensen, K. H. 1999. *Rush-hour Blues or the Whistle of Freedom? Understanding Modern Mobility WP 3/99*, Senter for Teknologi og Samfunn, Norges Teknisk-Vitenskapelige Universitet, Trondheim.

Sparke, M. 2006. Political geography: Political geographies of globalization (2)-governance. *Progress in Human Geography* 30(3): 357–373.

Sparke, M. B. 2006. A neoliberal nexus: Economy, security and the biopolitics of citizenship on the border. *Political Geography* 25: 151–180.

Sparks, J. R. and A. E. Bottoms 1995. Legitimacy and order in prisons. *British Journal of Sociology* 46(1): 45–62.

Sparks, R., A. Bottoms and W. Hay 1996. *Prisons and the Problem of Order.* Clarendon Press, Oxford.

Spelman, W. 2009. Crime, cash, and limited options: Explaining the prison boom. *Criminology & Public Policy* 8(1): 29–77.

Spens, I. 1994. A Simple Idea in Architecture. In *Architecture of Incarceration*, edited by I. Spens. London Academy Editions, London.

Squires, P. 2012. Neoliberal, brutish and short? Cities, inequalities and violences. In *Criminalisation and Advanced Marginality: Critically Exploring the Work of Loïc Wacquant*, edited by P. Squires and J. Lea. The Policy Press, Bristol, pp. 217–242.

Srikantiah, J. 2007. Perfect Victims and Real Survivors: The Iconic Victim in Domestic Human Trafficking Law. *Boston University Law Review* 87: 157–211.

Staeheli, L. A. and D. Mitchell 2007. Locating the public in research and practice. *Progress in Human Geography* 31(6): 792–811.

Standing, G. 2011. *The Precariat: The New Dangerous Class.* Bloomsbury, London.

Stevens, A. 2012. *Offender Rehabilitation and Therapeutic Communities: Enabling Change the TC Way.* Routledge, London.

Stoller, N. 2003. Space, place and movement as aspects of health care in three women's prisons. *Social Science and Medicine* 56: 2263–2275.

Stone, P. 2006. A dark tourism spectrum: Towards a typology of death and macabre related tourist sites, attractions and exhibitions. *Tourism* 54(2): 145–160.

Strange, C. and M. Kempa 2003. Shades of Dark Tourism: Alcatraz and Robben Island. *Annals of Tourism Research* 30(2): 386–405.

Stucky, T. D., K. Heimer and J. B. Lang 2005. Partisan politics, electoral competition and imprisonment: An analysis of states over time. *Criminology* 43(1): 211–248.

Svensson, B. and K. Svensson 2006. 'Inmates in motion – Metamorphosis as governmentality – a case of social logistics', Working Paper Series 5. University of Lund.

Swanson, K. 2013. Zero Tolerance in Latin America: Punitive Paradox in Urban Policy Mobilities. *Urban Geography* 34(7): 972–988.

Sykes, G. M. 1958. *The Society of Captives: A Study of a Maximum Security Prison*. Princeton, NJ: Princeton University Press.

Tartaro, C. 2003. Suicide and the Jail Environment an Evaluation of Three Types of Institutions, *Environment and Behavior* 35(5): 605–620.

Thies, J. 2000. Prisons and host communites: Debunking myths and building community relations. *Corrections Today* 62: 136–139.

Thies, J. 2001. The 'big house' in a small town: The economic and social impacts of a correctional facility on its host community. *Criminal Justice Studies* 14(2–3): 221–237.

Thompson, C. and A. B. Loper 2005. Adjustment patterns in incarcerated women: An analysis of differences based on sentence length. *Criminal Justice and Behavior* 32(6): 714–732.

Thompson, H. A. 2010. Why mass incarceration matters: Rethinking crisis, decline, and transformation in postwar American history. *The Journal of American History* 97(3): 703–734.

Thrift, N. 2000. Still life in nearly present time: The object of nature. *Body and Society* 6(3–4): 34–57.

Thrift, N. 2004. Intensities of feeling: Towards a spatial politics of affect. *Geografiska Annaler B* 86(B): 57–78.

Toch, H. 1977. *Living in Prison: The Ecology of Survival*. Free Press, New York.

Tootle, D. M. 2004. 'The Role of Prisons in Rural Development: Do they Contribute to Local Economies?' PhD Thesis, University of Georgia.

Touraut, C. 2009. Entre détenu figé et proches en mouvement. 'l'expérience carcérale élargie': Une épreuve de mobilité. *Recherches Familiales* 1: 81–88.

Touraut, C. 2012. Entre violence institutionnelle et régulation: Les effets de la prison au regard des proches de détenus, in Laforgue, D. and C. Rostaing (eds) *Violences et institutions. Réguler. Innover ou résister?* pp. 45–60.

Townsend, A. M. and J. T. Bennett 2003. Privacy, Technology and Conflict: Emerging Issues and Action in Workplace Privacy. *Journal of Labor Research* 24(2): 195–205.

Travis, J. 2005. *But They All Come Back: Facing the Challenges of Prisoner Re-entry.* The Urban Institute Press, Washington, DC.

Travis, J., A.L. Solomon and M. Waul 2001. *From Prison to Home: The Dimensions and Consequences of Prisoner Reentry.* The Urban Institute, Washington, DC.

Turner, B. 1995. Aging and Identity: Some Reflections on the Somatization of the Self. In *Images of Aging: Cultural Representations of Later Life*, edited by Featherstone, M. and A. Wernick. Routledge, London, pp. 245–262.

Turner, J. 2013a. Disciplinary Engagements with Prisons, Prisoners and the Penal System. *Geography Compass* 7(1): 35–45.

Turner, J. 2013b. The Politics of Carceral Space: Televising Prison Life. In *Carceral Spaces: Mobility and Agency in Imprisonment and Migrant Detention*, edited by D. Moran, N. Gill and D. Conlon. Ashgate, Farnham, pp. 219–238.

Turner, R. and D. Thayer 2003. 'Are prisons a sound economic strategy for New York? The views of rural policy makers', Paper presented at the meeting of the Northeast Political Science Association, Philadelphia, PA.

Turner, V. 1967. *The Forest of Symbols, Aspects of Ndembu Ritual.* Cornell University Press, Ithaca NY.

Turner, V. 1969. *The Ritual Process: Structure and Anti-structure.* Penguin, Harmondsworth.

Tye, C. S. and P. E. Mullen 2006. Mental disorders in female prisoners. *Australian and New Zealand Journal of Psychiatry* 40(3): 266–271.

Tyndall, A. 2010. 'It's a public, I reckon': Publicness and a Suburban Shopping Mall in Sydney's Southwest. *Geographical Research* 48(2): 123–136.

Ugelvik, T. 2011. The hidden food: Mealtime resistance and identity work in a Norwegian prison. *Punishment & Society* 13(1): 47–63.

Ugelvik, T. and J. Dullum (eds) 2012. *Penal Exceptionalism? Nordic Prison Policy and Practice.* Routledge, London.

Urry, J. 1990. *The Tourist Gaze.* London: Sage.

Urry, J. 2002. Mobility and Proximity. *Sociology* 36: 255–274.

Urry, J. 2006. Travelling Times. *European Journal of Communication* 21(3): 357–372.

Urry, J. 2007. *Mobilities.* Polity Press, Oxford.

Uteng, T. P. 2009. Gender, ethnicity, and constrained mobility: Insights into the resultant social exclusion. *Environment and Planning A* 41: 1055–1071.

Vagg, J. 1994. *Prison Systems: A Comparative Study of Accountability in England, France, Germany, and the Netherlands.* Clarendon, Oxford.

Van Gennep, A. 1960. *The Rites of Passage.* Chicago University Press, Chicago.

Van Hoven, B. and D. Sibley 2008. 'Just duck': The role of vision in the production of prison space. *Environment and Planning D: Society and Space* 26: 1001–1017.

Vanderburgh, D. J. T. 1992. The Hangman's New Clothes: Three Histories of Prison Reuse. In *Changing Places: Remaking Institutional Buildings*, edited by L. H. Schneekloth, M. F. Feuerstein and B.A. Campagna. White Pine Press, New York, pp. 137–150.

Vaz, P. and F. Bruno 2003. Types of self-surveillance: From abnormality to individuals 'at risk'. *Surveillance and Society* 3: 272–291.

Vidler, A. 1993. Spatial violence. *Assemblage* 20: 84–85.

Von Hofer, H. 2003. Prison populations as political constructs: The case of Finland, Holland and Sweden. *Journal of Scandinavian Studies in Criminology and Crime Prevention* 4(1): 21–38.

Wacquant, L. 2000. The new 'peculiar institution': On the prison as surrogate ghetto. *Theoretical Criminology* 4: 377–389.

Wacquant, L. 2002. The curious eclipse of prison ethnography in the age of mass incarceration. *Ethnography* 3(4): 371–397.

Wacquant, L. 2009. *Punishing the Poor: The Neoliberal Government of Social Insecurity.* Duke University Press, Durham, NC.

Wacquant, L. 2010a. Prisoner re-entry as myth and ceremony. *Dialectical Anthropology* 34: 605–620.

Wacquant, L. 2010b. Class, race and hyperincarceration in revanchist America. *Daedelus* 140: 74–90.

Wacquant, L. 2010c. Crafting the neoliberal state: Workfare, prisonfare and social insecurity. *Sociological Forum* 25: 197–220.

Wacquant, L. 2011a. The Wedding of Workfare and Prisonfare Revisited. *Social Justice* 38(1–2): 1–16.

Wacquant, L. 2011b. The great penal leap backward: Incarceration in America from Nixon to Clinton. In *The New Punitiveness. Trends, Theories, Perspectives*, edited by J. Pratt, D. Brown, M. Brown, et al. Routledge, Oxford, pp. 3–26.

Wahidin, A. 2006. Time and the prison experience. *Sociological Research Online* 11(1).

Wahidin, A. 2002. Reconfiguring older bodies in the prison time machine. *Journal of Aging and Identity* 7(3): 177–193.

Wahidin, A. 2004a. *Older Women and the Criminal Justice System: Running Out of Time*. Jessica Kingsley, London.

Wahidin, A. 2004b. Reclaiming agency: Managing aging bodies in prison. In *Old age and Agency*, edited by E. Tulle. NovaScience, New York, pp. 69–86.

Wahidin, A. and S. Tate 2005. Prison (E)scapes and Body Tropes: Older Women in the Prison Time Machine. *Body & Society* 11(2): 59–79.

Waitt, G., R. Figueroa and L. McGee 2007. Fissures in the rock: Rethinking pride and shame in the moral terrains of Uluru. *Transactions of the Institute of British Geographers* 32: 248–263.

Walby, K. and J. Piché 2011. The polysemy of punishment memorialisation: Dark tourism and Ontario's penal history museums. *Punishment & Society* 13(4): 451–472.

Wallerstein, I. 1998. The time of space and the space of time: The future of social science. *Political Geography* 17(1): 71–82.

Walmsley, R. 2013. *World Prison Population List* (10th edn). International Centre for Prison Studies, University of Essex.

Walters, W. 2008. Acts of demonstration: Mapping the territory of (non) citizenship. In *Acts of Citizenship*, edited by E. Isin and G. Nielsen. Zed Books, London, pp. 182–207.

Warren, J. I., S. Hurt, A. B. Loper and P. Chauhan 2004. Exploring prison adjustment among female inmates: Issues of measurement and prediction. *Criminal Justice and Behavior* 31(5): 624–645.

Watts, L. 2008. The art and craft of train travel. *Social and Cultural Geography* 9(6): 711–726.

Watts, L. and J. Urry 2008. Moving methods, travelling times. *Environment and Planning D: Society and Space* 26: 860–874.

Weiman, D. F. 2007. Barriers to Prisoners' Reentry into the Labor Market and the Social Costs of Recidivism. *Social Research* 74(2): 575–611.

Welch, M. and F. Turner 2007. Private Corrections, Financial Infrastructure, and Transportation: The New Geo-Economy of Shipping Prisoners. *Social Justice* 34(3/4): 56–77.

Wener, R.E. 2012. *The Environmental Psychology of Prisons and Jails: Creating Humane Spaces in Secure Settings.* Cambridge University Press, Cambridge.

Wheeler, S. 1961. Socialization in correctional communities. *American Sociological Review* 697–712.

White, A. A. 2008. Concept of Less Eligibility and the Social Function of Prison Violence in Class Societies. *The Buffalo Law Review* 56: 737.

Willett, J. and M. J. Deegan 2001. Liminality and Disability: Rites of Passage and Community in Hypermodern Society. *Disability Studies Quarterly* 21(3): 137–152.

Williams, A. 2002. Changing geographies of care: Employing the concept of therapeutic landscapes as a framework in examining home space. *Social Science & Medicine* 55(1): 141–154.

Williams, B. A., J. S. Goodwin, J. Baillargeon, et al. 2012. Addressing the Aging Crisis in U.S. Criminal Justice Health Care. *Journal of the American Geriatrics Society* 60: 1150–1156.

Williams, B. A., K. Lindquist, R. L. Sudore, et al. 2006. Being old and doing time: Functional impairment and adverse experiences of geriatric female prisoners. *Journal of the American Geriatrics Society* 54(4): 702–707.

Williams, E. J. 2008. 'A Tale of Two Prisons'. Doctoral dissertation, Rutgers, The State University of New Jersey.

Williams, N.H. 2007. Prison Health and the Health of the Public: Ties That Bind. *Journal of Correctional Health Care* 13(2): 80–92.

Willis, J. J. 2005. Transportation versus Imprisonment in Eighteenth- and Nineteenth-Century Britain: Penal Power, Liberty and the State. *Law & Society Review* 39(1): 171–210.

Wilson, A. 2008. *I Felt Seasick in the Van: The Movement and Mobility of Prisoners and the Effect on their Imprisonment.* Centre for Mobilities Research, Lancaster University Sociology Department.

Wilson, D., R. Spina and J. E. Canaan 2011. In praise of the carceral tour: Learning from the Grendon experience. *The Howard Journal of Criminal Justice* 50(4): 343–355.

Wilson, J. Z. 2008. *Prison: Cultural Memory and Dark Tourism.* Lang, New York.

Wilson, T. and J. A. Reid 1949. Malaria among prisoners of war in Siam ('F' Force). *Transactions of the Royal Society of Tropical Medicine and Hygiene* 43(3): 257–272.

Winchester, H. P. M., L. Kong, and K. Dunn. 2003. *Landscapes: Ways of Imagining the World.* Pearson, Harlow.

Wisnewski, J. J. 2000. Foucault and public autonomy. *Continental Philosophy Review* 33: 417–439.

Wolch, J., A. Rahimiam and P. Koegel 1993. Daily and periodical mobility patterns of the urban homeless. *Professional Geographer* 45(2): 159–169.

Wooldredge, J. 1991. Identifying Possible Sources of Inmate Crowding in U.S. Jails. *Journal of Quantitative Criminology* 7(4): 373–386.

Wortley, R. 2002. *Situational Prison Control: Crime Prevention in Correctional Institutions*. Cambridge University Press, Cambridge.

Wylie, J. W. 2007. *Landscape*. Routledge, London.

Yakubovich, V. 2006. Passive Recruitment in the Russian Urban Labor Market. *Work and Occupations* 33(3): 307–334.

Young, J. 2003. Searching for a New Criminology of Everyday Life: A Review of the 'Culture of Control'. *British Journal of Criminology* 43(1): 228–243.

Young, N. 2006. Distance as a hybrid actor in rural economies. *Journal of Rural Studies* 22(3): 253–266.

Yuill, C. 2007. The Body as Weapon: Bobby Sands and the Republican Hunger Strikes. *Sociological Research Online* 12(2) (2007): 1–11, http://www.socres online.org.uk/12/2/yuill.html, accessed 1/12/2011.

Zaitzow, B. H. 2011. We've Come a Long Way, Baby ... Or Have We? Challenges and Opportunities for Incarcerated Women to Overcome Reentry Barriers. In *Global Perspectives on Re-Entry*, edited by I. O. Ekunwe and R. S. Jones. Tampere University Press, Tampere, Finland, pp. 225–256.

Zamble, E. 1992. Behavior and adaptation in long-term prison inmates: Descriptive longitudinal results. *Criminal Justice and Behavior* 19(4): 409–425.

Zhang, Z., A. Spicer and P. Hancock 2008. Hyper-Organizational Space in the Work of JG Ballard. *Organization* 15(6): 889–910.

Zollo, S., M. Kienzle, P. Loeffelholz and S. Sebille 1999. Telemedicine to Iowa's Correctional Facilities: Initial Clinical Experience and Assessment of Program Costs. *Telemedicine Journal* 3: 291–301.

Zuelow, E. 2004. Enshrining Ireland's Nationalist History inside Prison Walls: The Restoration of Kilmainham Jail. *Éire-Ireland* 39(3 and 4): 180–201.

Index

A page number in **bold** denotes reference to an image.

abolitionism 5, 111, 151
activism 151
affect 17, 29, 41–2, 55, 113, 121–7, 147–8
Agamben, Giorgio 2, 17–20, 23–27
ageing 34, 45, 48, 50, 52–3, 77
agency 3, 17–28, 34, 40, 44, 50, 54, 72, 74, 83, 85, 92
Alcatraz 137
architecture, of prisons 4, 32, 44, 61, 113–24

backstage *see* frontstage and backstage
bare life *see homo sacer*
biopower 18–19, 25, 73
bordering
    borders 84

carceral habitus 38–9, 99–101
carceral tour 144
CCTV 121, 125
cell design 24, 120, 124, 127
cellular confinement 23, 30, 32–3, 40, 81, 93
census
    counting of prisoners in 64, 109
churning
    of prison populations 4, 105–6, 110
citizenship 23–4, 31, 33, 83, 100, 102
coercive mobility *see* mobility
conjugal visits *see* visiting prisoners
counter conduct 25–7
Crewe, Ben 11, 30–32
criminological cartography 150
criminology and criminologists 30, 43–4, 47–8, 54, 60, 113, 114, 115, 128, 147, 149–50
critical border studies 102

dark tourism 129, 137
de Certeau, Michel 2, 23, 28
dental care *see* teeth
Dirsuweit, Teresa 1, 3, 9, 21
dirty protest 38
disability
    disabled prisoners 76
disciplined mobility *see* mobility
disenfranchisement 64, 102, 109
distance 60, 66–70, 80, 83–4, 93, 150
docile bodies and docility 2, 3, 17–22, 26–7, 91, 94, 121

Eastern State Penitentiary 138–9, 141
electronic monitoring 92, 126
embodiment 4, 29, 34, 38, 40–41, 55, 92, 95–6, 99, 101, 127, 148, 149
emotional geography 29–35, 41, 125
environmental psychology 113, 119
exception (state of) 2, 6, 19–20, 23–4, 27
extraordinary rendition 73, 74, 84–5

family and effects of imprisonment; and separation 59, 68–9, 80, 84, 94, 96, 107, 142
femininity 31
food and eating 24, 33–4, 79
forced mobility *see* mobility
Foucault, Michel 1–3, 7, 8–9, 17–28, 47, 73–4, 88, 147
frontstage and backstage 30–33, 41
furlough 94–5

gender *see also* femininity; masculinity; transgender prisoners

gendered experience of imprisonment 27, 34, 40, 96, 99, 101, 108
Gilmore, Ruth Wilson 1, 4, 7, 65, 108, 110, 111, 147
Goffman, Erving 2, 3, 26–7, 30, 87–8
governmental mobility *see* mobility
Guantánamo 19
Gulag 67, 68, 129, 142

health
    health care 33, 40, 52, 67, 71, 77–8
        access to 33, 40, 52, 67, 71, 77–8
    healthy prisons 119
heterotopic
    heterotopia 4, 88–9
homo sacer 7, 17, 19–20, 23, 27
Hotel Lloyd 132
human rights 84, 122, 131
hunger strike 25–6, 38, 131, 135
hypermasculinity *see* masculinity

Joliet prison 135

Katajanokka prison 132, **133**
Kilmainham gaol 137, 139, 141

Langholmen prison 132, **133**, **134**
liminality 2, 82, 87, 89–91, 94, 101–2, 149
Long Kesh/The Maze 131, 135, 141

masculinity 31, 99, 100
Maze, The *see* Long Kesh/The Maze
media
    and imprisonment 61, 107, 108, 115, 129, 146, 147
    representations of prisoners 61, 107, 108, 115, 129, 146, 147
million dollar blocks 65, 151
mobility
    coercive, disciplined, punitive or governmental mobility 2, 3, 4, 10, 46, 71–86, 149, 151
    freedom of movement 2, 3, 4, 10, 46, 71–86, 149, 151

neoliberalism 13–14, 73, 106–8, 110
new punitiveness 4, 12, 13, 105–6, 110, 116, 120, 123
NIMBY 60–62, 69
noise 32

overcrowding 10, 32, 47, 68, 71, 120
Oxford prison 132

panopticon
    panopticism 8–9, 18, 20, 21, 22, 25, 27, 31, 73, 121, 127
Patarei prison 135, **136**
Patuxent prison 22
penal exceptionalism 115
Philo, Chris 1, 2, 10
Porridge 135
post-prison, the 105, 111, 129, 132–5, 145
power
    and prisons 8, 10, 18, 21, 25, 34, 50, 53, 72, 130
    theorisations of 8, 10, 18, 21, 25, 34, 50, 53, 72, 130
precariat 108
prison industrial complex 64
private prisons 63, 83
privacy, private space 21, 31–2, 41, 125, 126, 152
Prison Architect 145, 146
prisonfare 4, 7, 106–9
profit from punishment 70
punitive mobility *see* mobility
Punta Carretas prison 132, 134, 135

rehabilitation 5, 18, 21, 105, 110, 116, 122, 130, 139, 146
relational space 149
rendition *see* extraordinary rendition
reoffending 95–6, 106, 109, 116
Robben Island 131, 135, 137, 141

self-harm *see* suicide and self-harm
sentencing 106, 114, 116, 120, 132
sleep 38, 51, 127
solitary confinement 24, 122, 125, 144
sovereign power *see* Agamben, Giorgio

Stasi prison 81, **82**, 139, **140**
stigma 35–6, 96, 97–102, 107, 151
suicide and self-harm 118–21
supermax *see* solitary confinement
surveillance 2, 8–9, 17, 125–7, 147

tattoo 37, 97–8
teeth 35–7, 98–100

timespace 3, 28, 43–55, 151
total institution 2, 3, 26–7, 29, 34, 60, 87–9
transcarceral space 87, 92–7, 102
transgender prisoners 40

visiting prisoners 51, 68, 90–92

Wacquant, Loïc 4, 7, 11, 12, 28, 106–9

Printed and bound by CPI Group (UK) Ltd, Croydon, CR0 4YY

21/10/2024

01777088-0015